| DATE | | | |
|------|------|------|------|
|      |      |      |      |
|      |      |      |      |
|      |      |      |      |
|      |      |      |      |
|      |      |      |      |
|      |      |      |      |
|      |      |      |      |
|      |      |      |      |
|      |      |      |      |
|      |      |      |      |
|      |      |      |      |

4 50 5

# The
# Woodcutter's
# Companion

# The Woodcutter's Companion

## A Guide to Locating, Cutting, Transporting, and Storing Your Own Firewood

by **Maurice Cohen**

*Illustrated by Mark Schultz*

Rodale Press, Emmaus, Pennsylvania

Printed in the United States of America on recycled paper, containing a high percentage of de-inked fiber.

*Book design by Kim E. Morrow*

*Cover photograph by Scott Schmidt*

**Library of Congress Cataloging in Publication Data**

Cohen, Maurice, 1929–
    The woodcutter's companion.

    Bibliography: p.
    1. Fuelwood cutting. 2. Fuelwood. I. Title.
SD536.5.C63   634.9'82   80–26009
ISBN 0–87857–328–3 hardcover
ISBN 0–87857–329–1 paperback


2 4 6 8 10 9 7 5 3 1 hardcover
2 4 6 8 10 9 7 5 3 1 paperback

for my wife, Caroline

# Contents

Preventing Wood Rot
Damage to Dwellings from Wood Pests
Accessibility

# Introduction

This book has been written for the amateur woodcutter
—someone who cuts firewood for economy, health, and the
pleasures of working out-of-doors. I am an amateur woodcutter. I have never logged or done tree work for a living;
I have never sold firewood. I wasn't raised on a farm or in
wooded foothills. During a good part of my working life I
have been a teacher, pursuing my sedentary occupation in
the cities or suburbs. But if I have never *sold* firewood,
neither have I *bought* any during many years of providing
wood for my family's fireplaces. I have enjoyed the best kind
of exercise. And, though I have done a good bit of hiking and
camping, some of my happiest memories of woods and open
places come from woodcutting.

This book has been written, then, as much to share the
pleasures of woodcutting as to provide the information you
need to safely cut firewood for fuel. Properly seasoned wood
is good fuel. A ton (64 cubic feet) of well-dried hardwood is
the heating equivalent of ½ ton of coal, 75 gallons of fuel oil,

or 12,000 cubic feet of natural gas. It burns cleanly, producing wood ashes as a by-product of the combustion process—and these can be used as garden fertilizer. As more and more people have turned to wood as a primary or supplementary fuel, the cost of ready-cut firewood has continued to rise. But large amounts of burnable wood are available free for the asking, often within city limits, to someone who has learned the art of woodcutting.

Your decision to become a woodcutter may reflect a determination to become self-sufficient and live in a simpler, more frugal way. This means buying only what is necessary in the way of clothing, tools, and equipment. I have done my best to keep your shopping lists as slim as possible. But if you really want to save, you'll have to do your part by buying only what you need. This may not be as easy as it sounds, once woodcutting becomes part of your life.

The economies of cutting your own firewood are plain enough; not quite so well known are the physical benefits. Woodcutting is a total exercise, particularly when you do it with hand tools. Every part of the body comes into play as you build woodcutting into your daily life during autumn, winter, and early spring.

Though woodcutting itself is best confined to the cooler months of the year, there are other activities, some of them quite closely related, which you can engage in to keep in shape during the warmer months. You'll find that you will enjoy hiking and camping more once you've learned to use an axe skillfully. Gardeners will find that breaking clods with a mattock, deep raking, digging, hoeing, and pushing a wheelbarrow use the same muscles and develop skills like those used in hand woodcutting. (These are only two ways of staying in condition outside of the woodcutting season; I'll be mentioning others.)

Important as the economic and physical benefits of woodcutting are, you may find that its greatest benefit is its effect on your frame of mind. Hand woodcutting can be an uplifting experience. You learn to work with the wood, rather than trying to force your will on it. This lends you serenity, as well as providing you with an opportunity to continually develop your woodcutting skills.

I am convinced that this is an opportunity for self-knowledge that the beginning woodcutter should not miss, so don't start out by acquiring a chain saw. Get the basic hand woodcutting tools, which you will need anyway, and learn to use them first. Then, if you decide you need it, get a chain saw. Proceeding this way, you'll purchase more wisely, work more safely, and get much more satisfaction from your woodcutting.

I've said something now about the benefits and pleasures of woodcutting. Let me also caution you that woodcutting can be very dangerous if you rush into it without knowing exactly what you're doing. There is *no* safe substitute for proceeding step by step: acquiring skills progressively; conditioning yourself gradually; and learning to judge at every stage exactly what your personal capabilities are. This doesn't mean you can't bring home a good deal of firewood safely your first time out! You just have to know where to go and what to tackle initially.

I have spent quite a few years working in the woods without injury. I intend to keep my record clean; you can do the same if you read this book carefully and absorb its contents before you go off to do your own cutting. Read the *whole* book. There is no separate chapter on safety, because safety is the most important consideration in every wood-gathering and cutting activity. Keep this in mind while you limb a fallen oak, or quarter and split basswood in your backyard, and you'll feel a sense of deep satisfaction and no regrets when you watch the wood burn in your fireplace.

# Chapter 1

# The Art of
# Reconnoitering
# for Firewood

If you own a large wooded property and can cut your own trees, it is no problem to locate firewood. But what if you live in the suburbs, or in a city? Is it still possible to keep your stove or fireplace stoked with free firewood?

I can't promise anyone 6 cords of free wood each year, but despite the renewed interest in wood as a fuel, there is still a lot of wood out there for the taking. As a matter of fact, more wood is probably discarded than salvaged.

## Where There's Wood
## for the Taking

When I began cutting firewood, we were renting a house on 100 wooded acres in Westchester County, New York. We then moved to Nyack, a village on the Hudson about 20 miles north of New York City. Now we live in

*1*

downtown Charleston, South Carolina. I would have thought that as we de-rusticated, it would be necessary to travel farther and farther away from home for wood, but this has not been the case. This year, after a winter of daily fires in one or more open fireplaces, I am ending the heating season with a fair supply of seasoned firewood left over—and much of it was collected no more than a few blocks from our house.

How much wood you can collect and how far you have to go to get it will depend on where you live, but only to a certain extent. A good deal depends on how thoroughly you survey all the possible sources of wood in your area. To give as complete a picture as possible, I'll assume that you, too, are living in a built-up area. Think in terms of a number of concentric zones, with your home at the center:

1. *Your immediate neighborhood* —the lot you're living on and those on the streets around it, within a radius of five or ten blocks

2. *The city or town where you live* —including parks and the local sanitary landfill (the solid waste "dump")

3. *Nearer and outer suburbs*

4. *The countryside* —the area up to about an hour's driving time from your house

In my own case, for example, there is our immediate neighborhood in downtown Charleston; peninsular Charleston, including Hampton Park and the municipal landfill; the suburbs of Charleston; and the countryside within a radius of 30 or 40 miles. This last zone includes the Francis Marion National Forest, a Westvaco Lumber Company forest, and other large private wooded areas (see Figure 1-1). Each of the four zones has its own wood collecting possibili-

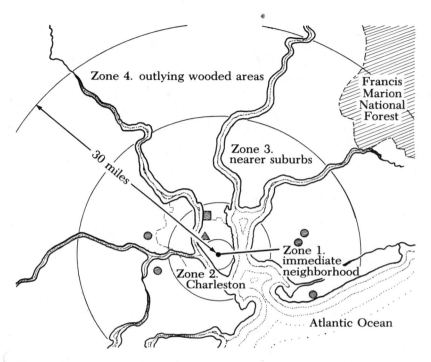

**Figure 1-1:** Within 30 miles of the city of Charleston, South Carolina, there are many sources of wood.

ties, and for each there are specific ways of obtaining information and permission.

# The Big Woods

State, federal, and commercially owned wooded properties are likely to be found in Zone 4. For federal forests, simply call your nearest forest ranger. You can find the phone number in the telephone directory under United States Government, Department of Agriculture, Forest Service. For state forests, look up your state forestry commission. It's also a good idea to visit your local soil conservation office. The people there know local foresters personally. They can also tell you a good deal about the management and access policies of owners of large wooded properties, including the forests of timber and pulpwood producers. From my local soil conservation agent, I found out that Westvaco permits private individuals to salvage and cut wood on its property in the Charleston area after commercial harvesting operations are completed.

There are, of course, other ways of getting such information about company practices. The local newspaper will have articles on woodgathering; tree service people who service large private holdings may be willing to provide information; and large commercial owners can and should be called directly. But if you want to do more than just get free wood as quickly as possible, be sure to do at least a bit of visiting with local forestry and soil conservation people.

People who do this kind of work are usually eager to share their information and contacts. Federal soil conservation people know state foresters, state foresters know their private counterparts, and so on. If you visit the federal and state offices, you may also find that they have available for free distribution publications on the use and care of trees, land management practices, and other matters that will concern you as you begin observing trees and what happens to them.

Another important source of information about wood in the outermost zone is your state highway department. Road building and widening operations often require the removal of trees, and even when commercial woodcutters subcontract the work and remove the best wood, there can be more than enough left over for individuals who know how to work the less easy material and are willing to walk a few yards from the road.

Soon after we moved to Charleston, I learned that permits are granted for free firewood cutting in Francis Marion Forest, which begins about 30 miles from my house. I have yet to get that far. The year we arrived, extensive road work was going on about 10 miles out of town. I called the highway department and was told I could take all the cut wood I wanted. Huge piles of wood could be seen for miles along the road being widened. What other woodcutters and I did not take was burned on the spot as waste.

Since that winter in Westchester more than ten years ago, I have not cut wood on private tracts. But given the fact that many people who own wooded land don't have the time or inclination to do the kind of work that goes with good small forest management practices, there should be room for cooperation between landowners and knowledgeable amateur woodcutters.

If my other sources dried up, I would certainly explore that area. I would talk to local foresters and ask them for suggestions about likely prospects. State forestry personnel will provide, on request, advice on tree and forest management to property owners. This means that at no cost to the property owner an expert can help him or her evaluate needs and interests and make specific suggestions. I could thus assure the property owner that cutting would be done only after a plan, tailored to the owner's goals, had been developed under the guidance of a trained forester.

One of the lesser quandaries of our time is why more property owners don't make use of such help. In many cases, the result would be safer, more stable, and attractive property—and a continuous supply of firewood from pruning and beneficial thinning.

# Suburban Wooded Sites

Unless you live in the suburban zone, or have friends there, your most likely source of firewood in Zone 3 will be demolition or construction sites. If you have any doubt at all about whether it's legal to take wood from a spot, ask the people working there. If it's a weekend, or the site looks inactive, ask the local police.

Outlying sites can be nice places to work in—particularly if they're abandoned! The first such place I worked in New York was a "cut" off a side road a mile or so outside Nyack. A number of red oaks had been toppled by a bulldozer and pushed around a bit. I checked with a local policeman before doing any cutting. He told me that the place had been that way for a long time, and he saw no reason why I couldn't go ahead and cut firewood there. I worked there on and off all that winter and never exhausted that one site (other trees there had toppled because of erosion). It was a lovely spot, sheltered by a wooded hill, and with no dwellings in sight. One corner was marshy, but it froze solid and that made it easy to drag logs to my roadside parking spot.

In the Charleston area, I found no abandoned sites of that size and quality. But one winter I got all the live oak I could carry from a heap of trees left for months by some "developer." The year before last, I got a few vanloads of wood from a lot that was being cleared for a fast-food restaurant. The easiest stuff to get at had already been taken by the chain saw boys, but there was plenty of fuel left for anyone willing to clamber around a bit. There was certainly no romance about that spot, only the somewhat melancholy satisfaction of being able to cremate honorably some of the remains of half an acre of old beech and oak.

# Woodgathering within the City Limits

For woodgathering within the innermost zones, it is important to be familiar with the subdivision of forestry known as *urban forestry.* Charleston County, for example, has an urban forester, a state employee whose office is well within the city limits. I first learned of his existence from a fairly lengthy interview published in the newspaper. It's a good thing that I had a chance to find out in this fashion that urban forestry has a great deal to tell anyone, not just the amateur woodcutter on the prowl for wood. If I had heard just the title, "urban forester," I suspect I would have been confused.

My last contact with trained foresters had been in the foothills of western Alberta, Canada. They lived in the midst of moose, elk, mule deer, mountain sheep, black and grizzly bears; their children went to school with Indian children. It's wonderful to meet such people in their "exotic" work places. But if you live in a built-up area and want to cut your own firewood, don't neglect visiting your urban forester and talking with him about *his* work place—the total ecological unit in which you live. It's a splendid way to get not only an informed survey of the wood available in your entire urban/suburban region, but also information about what needs to be done for trees in our cities and suburbs.

Within the inner zones, investigate the sanitary landfill, municipal parks, and power companies as possible sources of wood. In Charleston, the landfill is open to the public every Saturday morning for woodcutting and lumber salvaging. There are often a few firewood gatherers with chain saws and pickup trucks, but they make little difference to the acres of trunks and boughs that remain to be shoveled under by the bulldozers during the week. The main problem with going to the landfill to look for firewood is that the scrap lumber (from crates and construction sites) to be gotten there for the taking is usually more tempting!

Private individuals cannot cut wood within the boundaries of public parks, but the park crews have the problem of disposing of material they remove in the course of pruning and thinning. In Charleston, park personnel select what can be used in the park's own buildings, then leave the rest— which can be a sizable amount—in a nearby lot for a while. Private individuals can cut on the lot if they wish; what is left is then taken to the landfill.

I have not gotten wood from power company crews, though I know they cannot keep up with the pruning that needs to be done to protect overhead power lines. I have

seen a lot of wood that could have been first-rate firewood
going into their chippers.

I did get a vanload of fine wood from a municipal crew
working around the corner from my house. On request, they
will prune trees on the street, if there is any hazard to houses
or pedestrians. Most of the time they, too, feed what they cut
into the chipper, but when I asked the crew if I could have
the large boughs they were cutting, they were more than
agreeable. One of the workers is in the habit of setting aside
wood for making tables and birdhouses, and to give to needy
elderly people. When I came back to my vehicle to collect
the green live oak, I found the wood neatly cut and stacked
and ready for me to haul away.

# Hunting Wood at Home

I think I could get all the wood I could ever want from
the sources just mentioned in my urban zone—the landfill,
leavings by park personnel, and other cutting by work crews
in the city. But I have found another source, so close to my
house that this year, even though I collected easily as much
wood as I have ever done, I did not have to drive any distance worth mentioning to pick it up; in fact, I conveyed a
good bit of it in a wheelbarrow.

Since I come from a family in the habit of walking, I
have been perambulating my neighborhood in Charleston
since we arrived here. I have gotten to know intimately the
streets, houses, yards, and *trees* in a radius of at least ten
blocks from where we live.

Time and time again, I have seen eminently burnable
wood, sometimes in substantial amounts, left out on the sidewalk to be picked up by the city sanitation crews. When I
have the time, I pick it up first. Because I hate to see firewood wasted, I'm troubled by the degree to which this part
of my system is still inefficient and inelegant. I really need
to be carrying an axe and wearing a pack frame on better

(richer-in-wood) days as I walk my daughter to school. I don't think she would mind, and I'm sure her friend, whom we pick up on the way, would be supportive, since she now makes a point of telling me when she sees "good wood" left out on the street!

Then there are the storms, which can be very severe here. I don't need them as a source of wood, and I wish they didn't occur, but after a bad storm huge quantities of wood have to be disposed of by the city workers. I collected more than a cord of wood in my immediate neighborhood after the last major storm, and could have gathered much more if I had needed it. Most of it was live oak and magnolia, both of which make excellent firewood when seasoned.

Much of the tree damage that produces this literal windfall is avoidable. It results from the extent to which trees are neglected in cities. Pruning, bracing, and other techniques would save many trees and prevent a lot of the damage that occurs to property during storms. Sometimes I have fantasies of approaching *urban* property owners and offering to help them care for their trees in return for the wood that needs to be cut to keep the trees healthy and safe.

# Making the Method Work for You

This completes my presentation of The Method. Of course, it needs modification by others who may want to use it. But the principles are valid. Where there are trees, there is the possibility of firewood. If trees are not pruned, cared for, and thinned or removed deliberately, sooner or later a lot of wood is going to be brought down by natural causes.

Don't let it go to waste. Take advantage of the free wood that is available to you. But learn to be choosy. Some kinds of wood make better, *safer* fires than others. Prepare for your forays in search of wood by reading the next chapter, "What to Burn."

# Chapter 2

# What to Burn

As the Spirits of Darkness be stronger in the dark, so good Spirits which be Angels of Light, are augmented not only by the Divine Light of the Sun, but also by our common Wood Fire; and as the Celestial Fire drives away dark spirits so also this our Fire of Wood doth the same.

—Cornelius Agrippa

It's one thing to locate wood, another to decide what is worth cutting and taking home for fuel. Once you find a mass of wood available for the taking, how do you tell whether it is worth working on?

To answer, back off a bit and consider, not the stuff immediately before you, but what you need eventually for your stove and fireplace. Then you can look at the tangle of logs, trunks, and limbs in front of you and decide how much of it justifies the time and labor you will have to put in to turn it into fuel.

# The Character of Good Fuel Wood

Without getting too technical, it is easy to state the desirable characteristics of fuel wood: first, it should be *dry*. You can burn wood that is wet or unseasoned, but then you waste for heating purposes the heat energy needed to dry out the wood in the fire. Practically, this means that green wood—wood from trees that were alive just a little while ago —has to be seasoned, or dried in the open air for a minimum of six months, in order to get it dry enough to burn with reasonable efficiency (see Chapter 7, "Transporting and Storing Firewood," for more on seasoning).

Second, the *heavier* the wood is after seasoning, the more heat you can get from a given volume of it, and the slower it burns. Once it is dry, any wood will burn more or less readily, even wood so decayed that it falls apart if you bang it against something. Light, dry wood is porous. The lighter wood is, the quicker it burns and the more of it you must keep on hand for a long fire.

This isn't always a disadvantage. If you want to get the chill out of a room quickly, you want a fire that you can start quickly and which will put out a lot of heat fast. Later on, you will want a steady, slow-burning fire. To meet both these needs, stock seasoned lighter wood, as well as denser wood.

Fuel wood should not have objectionable characteristics. It should not have been treated with chemicals; it shouldn't "spit" (throw out live coals) when it burns; and it should not be excessively flammable because of a super-abundance of resin.

You can avoid chemicals by not using posts, timbers, and treated construction lumber for fuel. Railroad ties and wood that has obviously been treated with creosote should never be used at all. In general it's best to pass by wood from docks and outdoor structures, as well. Even if you can't smell the chemicals, the wood has probably been impregnated with something that makes it unsuitable as fuel.

Spitting is a property of more than a few species of

woods. The lighter woods tend to be more of a problem, though some of the best heavy hardwood fuels can throw out small, but dangerously long-lived, coals. Hemlock and spruce, which are often used in carpentry and construction, are notorious for spitting.

It is not economical, however, to dispense with all the woods that spit. With the good hardwood fuels that may spit a bit occasionally, just make it a rule always to observe the basic precautions for an open fireplace: a good fire screen, and nothing flammable on the floor in front of the fireplace. Be careful, too, about wool rugs. They don't ignite readily, but they can smolder for a long time from a spark.

I wouldn't use the lighter woods that spit in any quantity in an open fireplace, even with a good screen. The popping and fireworks display are not relaxing. But the evergreens have other disadvantages, too; because of the resins they produce, in large quantities they burn too quickly and uncontrollably. Burning a Christmas tree, green or dry, can lead to a chimney fire if there is any creosote built up inside the flue or stove pipe. In very small amounts, however, spruce or balsam fir make excellent kindling. You can store them outside, and they will burn like the dickens even when wet, so it is well worth keeping some on hand.

Finally, your fuel wood should be what is available to you. Use what you can; don't worry about what is ideal. Spruce, as I have indicated, is not an ideal fuel, but when wood was the only fuel available to white settlers in Alaska, they survived and thrived, burning spruce all winter in their stoves. Northern poplar isn't ideal fuel wood, either, but Canadian Indians in Alberta, where the winters are subarctic, make their cabins cozy with stoves stoked with poplar.

# Figuring Fuel Efficiency

Generally, then, you want wood that is dry, dense, and available. Is that all? What about the relative advantages of this or that type of wood? How about those awesome Btus? On the Btu rating charts, black locust and shagbark hickory are winners, white pine and basswood are losers. Shouldn't you know one from the other?

It is good to be able to recognize different trees and their woods, in winter as well as in summer. I think, however, that the benefits are more esthetic and spiritual than practical. That doesn't make the knowledge any less valuable! Indeed, I'd much rather see someone studying trees because trees are friendly and nice to know for themselves, than because some are "good" and others "bad" on a fuel efficiency scale.

But let's go back to Btus for a bit so I can explain how the heating capacity of wood is calculated. A Btu (British thermal unit) is a measure of heat. To be exact, it is the quantity of heat needed to raise the temperature of 1 pound of water through 1 degree Fahrenheit; it's equal to 252 calories. Btu counts are not based on actually heating pots of water with wood from somebody's woodpile. If they are based at all on actual combustion experiments, you can be sure that the work was done in a laboratory with selected samples dried to a uniform moisture content.

The basic chemistry of wood doesn't vary from species to species, and has been known since the nineteenth century. Chemists have known for more than 100 years how to calculate theoretically the amount of heat that should be obtainable from burning a given quantity of dry wood.

Broadly speaking, tables comparing heat values of different woods are compiled in two ways. In the first method, the weight per cubic foot is multiplied by a constant (usually 80) to get a figure for the number of pounds per cord. The estimated number of pounds in each cord is then multiplied by another constant (usually in the neighborhood of 7,000) to get a figure for the number of Btus per cord. Using figures for density or pounds per cubic foot provided by "Fuelwood and Woodburning Stoves," a Pennsylvania Cooperative Extension Service publication, we can calculate for basswood,

a)   24.8 pounds per cubic foot × 80 (estimated number of pounds
     per cord) = 1,984 pounds per cord;

b)   1,984 × 6,900 (Btus per pound at 20% moisture content) =
     13.8 million Btus per cord.

Sometimes a single Btu constant is not used, but different constants are used for resinous and nonresinous woods.
Larry Gay, in *The Complete Book of Heating with Wood*
uses 7,000 as his Btu constant for hardwood, but 7,360 for
Douglas fir and 7,200 for white pine, "to take into account
the inflammable resins in the latter two."
     Another method of compiling heat value tables for diff-
erent woods is to work with chemical analyses. The results
of the analysis by weight of the combustible elements—
carbon, hydrogen, and oxygen—in each type of wood are
then inserted into a standard equation to give a figure for
the Btus per pound. This method gives much higher Btu
figures for resinous woods.
     To understand the actual compilation of a Btu table,
take a look at the table on the facing page, which my son
made with the help of a computer at a nearby college.
     If, then, you want to compute theoretically the Btus for
a given type of wood, all you need are a figure for its density,
or weight per unit volume, and a number for the Btus in *any*
pound of dry wood. What all this amounts to is that the
heavier the wood after it has been thoroughly dried, the
more Btus it will be assigned, because there is more wood
to burn in a given volume of it.
     A long time ago, a chap named Rumford calculated
theoretically that a pound of wood "perfectly dried by artificial heat" should deliver 6,582 Btus, and that a pound of
wood "in ordinary state of dryness" should give 5,265 Btus;
actual experiments gave 6,480 and 5,040 Btus respectively.
I got Rumford's figures from the ninth edition of Thomas
Box's *Practical Treatise on Heat as Applied to the Useful
Arts, for the Use of Engineers, Architects, Etc.,* which was
hot off the press in 1900. The first edition of this book goes
back to 1868.

| Wood | Density[a] | Pounds per cord[b] | Btus per cord (millions)[c,d] |
|------|-----------|-------------------|------------------------------|
| apple | 48.5 | 3,880 | 27.1 |
| aspen | 27 | 2,160 | 15.1 |
| basswood | 26 | 2,080 | 14.5 |
| beech | 44.3 | 3,544 | 24.8 |
| dogwood | 51.5 | 4,120 | 28.8 |
| black locust | 49 | 3,920 | 27.4 |
| silver maple | 33.9 | 2,712 | 18.9 |
| white oak | 46.8 | 3,744 | 26.2 |
| sweet gum | 36.4 | 2,912 | 20.3 |
| white cedar | 23.8 | 1,904 | 13.2 (13.6) |
| white pine | 25.4 | 2,032 | 14.2 (14.7) |
| longleaf pine | 41.6 | 3,328 | 23.2 (24) |
| black spruce | 28.8 | 2,304 | 16.2 (16.7) |

[a] The density figures (weight per cubic foot) are those given in *The Wood Handbook* in Table 7, and are for wood at 15% moisture content.

[b] The figure for pounds per cord was obtained by multiplying the density by 80 (a cord is 128 cubic feet by volume; this allows for the spaces between the wood).

[c] The Btu figures were obtained by multiplying the pounds per cord by 7,000.

[d] The last four woods are coniferous. To give a rough adjustment for the presence of resin, an additional heat value has been given in parentheses; this results from adding another 3½% to the figure arrived at with the basic constant of 7,000.

If you feel that knowing the density of this or that wood brings in another mystery, don't worry. Your hands and your axe are pretty good density meters, particularly when wood is dry. Numbers for the density of different kinds of wood are also easy to come by. The source I like best is *The Wood Handbook*, published by the Forest Service of the United States Department of Agriculture. If you really get interested in wood, don't be surprised if this handy volume, which is still a bargain, gets on your bedtime reading list.

This book contains a table called "Strength Properties

of Some Commercially Important Woods Grown in the United States." Almost 100 types of wood are analyzed, with sixteen columns of figures for each wood. There is poetry in those numbers, so great is the richness of the information they contain. You won't find Btus or even density listed, but two specific gravity figures are given for each wood, one for when it is green, the other for when it is seasoned. Specific gravity and density are so closely related that they are really only two ways of talking about the same property, the specific gravity for a substance being no more than a number comparing its density with that of water. This is why a wood like black locust, with a relatively high specific gravity (0.69 when dry) is a Btu star, while poor little old white pine, with a shameful specific gravity of 0.35, is ready to fall off the fuel efficiency scale.

With a copy of *The Wood Handbook*, you, too, can be an expert and compare in detail the fuel efficiency of 100 different kinds of wood. But the simple truth is that for home heating purposes, you don't need the numbers. It's the dry weight of the wood that counts, no matter what wood you're concerned with. The heavier the wood after it has been thoroughly dried, the more heat it can provide. When you are looking through the seasoned wood in your woodpile for firewood, compare weights by heft to choose wood to meet your needs.

In other words, don't get carried away by the Btu numbers. What really matters is how much time, money, and effort you have to expend to get your fire going. And nobody but you can figure *that* out. The figures say you need almost twice as much basswood as black locust to get the same amount of heat out of a stove or fireplace. Theoretically, then, black locust has close to twice the fuel efficiency of basswood. But what if you can get hold of a generous quantity of basswood, close to where you live, for nothing, but don't have a ghost of an idea where to find a black locust you can have for fuel? Availability is a more important consideration, in this case, than fuel efficiency.

In deciding what to burn, difficulty arises when the most available wood is resinous, from the conebearing or evergreen trees. The practical problem for the home wood fuel user is not whether the people who assign high Btu ratings to such woods are theoretically correct, but whether it is safe to burn such woods at home in large quantities.

Since there can be a relationship between creosote formation and resin content, you can see what the most prudent initial answer has to be. Don't burn resinous woods in any quantity until and unless you have discussed the matter with knowledgeable local individuals. Agriculture extension personnel and foresters can be counted upon to be helpful. Don't overlook old settlers and experienced masons, plumbers, and contractors. Fire-fighters are particularly anxious to prevent problems before they arise, since they know only too well what happens when avoidable mistakes are made.

Trees are variable creatures. Members of the same species differ from region to region, and there are many types of pine alone. Eastern white pine, for example, is lighter than every type of oak; longleaf pine is almost as dense as oak. Resin content varies greatly, too. Fortunately, there is no problem in knowing what is resinous. You don't have to know the difference between loblolly, longleaf, shortleaf, and slash pine, or even between the spruces and the firs to know when there is resin in the wood you're handling. It sticks to your hands and tools and has a rich aroma.

# Recognizing Good Fuel Wood in Summer

An awful lot of good people have grown up in the last generation or so without much more knowledge about tree identification than a vague notion that there is a difference between "evergreens" and trees that drop their leaves in the fall. They may not want to play the Btu numbers game, but they want to know at least what they are comparing with what, and they don't have time to take night courses in botany.

Generous help in this respect is no further away than the local public library, which has excellent books on tree identification. In addition to the standard general works, it will also have books dealing with the particular region in which you live.

There are many truly fine guides to tree identification. Without pretending to give a complete list, or suggesting that works not listed aren't just as good as the ones cited, I'd like to mention a few that are likely to be on your library's shelves. These books assume no previous knowledge of woodlore, have good illustrations, and can be scanned quickly to identify a particular tree. *A Field Guide to Trees and Shrubs,* by George A. Petrides, covers the northeastern and north-central United States and southeastern and south-central Canada. This easy-to-use volume contains a lot of information about the uses of wood, and the relationship of trees and wildlife.

William M. Harlow's *Trees of the Eastern & Central United States & Canada* is not as detailed as Petrides, but is more compact and very readable. It, too, can teach you a lot about the uses of wood from different trees.

Canadian and northern United States readers—western as well as eastern—may want to know about a nontechnical Canadian work put out by the Canadian Department of Forestry and Rural Development. *Native Trees of Canada* is illustrated with large photographs of leaves, twigs, fruits, and bark to help in tree identification.

Every region has its own superb studies—and when it comes to tree books, age is not a drawback. For example, the whole of C. H. Greenwood's *Trees of the South* (published in 1939) is worth reading; her informal, anecdotal style covers gracefully a mine of information. Elwood S. Harrar and J. George Harrar's *Guide to Southern Trees,* a pocket manual with unusually clear and helpful drawings, is likely to be more readily available since it is more recent.

All of these books are easy to use, and so cannily designed for the botanically innocent that if it is mere identification you want, a rush through the illustrations alone will often give a fair idea of what you are trying to identify.

# Winter Botany

The emphasis in the books mentioned so far is on identification when you know the leaves, fruits, and general appearance of the tree in summer. What happens if you get curious in winter, when so many trees lose their leaves? Actually, if you're willing to study them a bit, the books just listed can be quite helpful—and there may still be some leaves around, withered on branches, or underneath a tree you're interested in identifying.

But there is also more specialized help, for there are books available on winter botany. Although Annie Oakes Huntington's *Studies of Trees in Winter* is out of print, you may find it worth looking up in the library or a rare book store. Published in 1902, with an introduction by Charles S. Sargent, author of the magisterial two-volume *Manual of the Trees of North America,* this book is authoritative as well as beautiful. The old-fashioned photographs are as evocative as Wyeth paintings; they also clearly show bark patterns and typical tree contours. Huntington was wonderfully literate. Browse through this book, and you will not only become familiar with northeastern trees in their winter undress; you'll also come away knowing a bit of history. Here, for example, is the author's explanation of why sycamores used to be called *grief* trees.

It is a favorite Scotch tree and was much planted about old estates in Scotland. Over two hundred years ago, the powerful barons in the West of Scotland used these sycamores for hanging their enemies and refractory vassals on, and these trees were called dool, or grief trees.

And did you know this about the gray and stately beech tree?

> Passienus Crispus, the orator, who married the Empress
> Agrippina, was so fond of it that "he not only delighted
> to repose beneath its shade, but he frequently poured
> wine on its roots, and used often to embrace it."

A more sober and analytical companion to *Studies of Trees in Winter* is William Trelease's *Winter Botany: An Identification Guide to Native and Cultivated Trees and Shrubs.* With the help of a magnifying glass, you can learn from Trelease how to identify a tree or shrub by its general shape, the appearance of its winter buds, and the structure of the twigs in cross section. This book, too, is written for the layman, with a helpful glossary in the back.

With resources like these books available, it isn't at all hard to get to know a good bit about trees. An hour or two looking at pictures will get you started. You can begin your course in dendrology (the scientific study of trees) in the winter as well as in the summer. You will discover that even a little knowledge of the identity and character of different types of trees can be a large gain for the woodcutter.

# Seeing the Forest for the Trees

Where you live, work, walk, and look for wood is both special and significant. Every region of the continent has its differences from every other; there are contrasting ecological zones within each geographic region; and different habitats and subhabitats in each zone. Trees are not simply dumb, woody things; each type has special needs, and if those needs aren't satisfied, the tree won't grow. In many parts of North America, a drive of only a few hours north or south will take you through several ecological zones, each with its predominant tree types—or, more exactly, *com-*

*munities* of trees. Like people, trees socialize, and not only with members of the same species.

If you live near mountains, you don't have to go far horizontally; going *up* can produce even more dramatic effects.

> Each 1,000 feet in ascent of a mountain is equivalent to a trip of 300 miles northward; that is why the top of a high mountain on the equator contains plants and animals similar to those found in the Arctic lowlands.
>
> —Farb, *The Forest*

The higher the ascent, the more dramatic the difference will be. Once I accompanied a snuff-chewing, mountain sheep hunter up one of the foothills of the Canadian Rockies. An hour of climbing took us from lush, temperate swampland to alpine meadows where chilly winds blew in midsummer. Way down below, everything looked like the green bottomlands and forests of New York state; past the tree line, however, there were unfamiliar flowers and strange, twisted little trees. It looked like Switzerland or the Arctic tundra—and it *was* like them.

*The Forest* is an example of another kind of good reading about trees. It provides an introduction to the unbreakable web of relationships into which we and trees and the rest of life on earth are woven. There are many other good books along the same lines, both general and more specific. Vinson Brown's *Reading the Woods* is a field manual for the layperson which describes in detail how to perceive the various types of forests as total environments.

# The Democratic Woodpile

By now you've probably realized that I would rather not give detailed advice on the superiority of one type of fuel wood over another. Sure, "other things being equal," oak is better than sycamore for fuel, and poplar is better than pine. Thoroughly rotten wood isn't worth the handling, and chemically treated wood should be avoided except in very small quantities, and then used only for kindling.

I believe in making the best of the wood I can get without unnecessary expense and unenjoyable labor. In practice, this means being willing to trim some of the decayed wood away from the otherwise sound wood in a log that has been lying on the ground for a while, and picking up boughs, branches, and sometimes even twigs that would otherwise be left as litter. In other words, I do everything sensible to collect and use adequate as well as noble wood. I want my woodpile to be like a genuinely democratic human community, with plenty of room for different types, including what is less than perfect.

There is a place for softwood as well as for hardwood, for small stuff as well as intermediate and large pieces. Some wood, like balsam fir, that I would not burn as large pieces, is worth keeping on hand for kindling; the same is true of odds and ends of construction lumber. Even big old pieces a little porous from decay are worth drying out if they aren't too crumbly. They don't make really good fuel, but will burn cheerily enough in the dead of winter. They're like toast left over from breakfast: not for serving to anybody else, but fine when you're snacking alone.

When you go out woodgathering, know enough to recognize good wood *and* to make the most of whatever wood is available to you. Learn from the resources at your local library, and the advice of your local forestry agent and other regional experts how to use the wood available to you intelligently. Whether you decide you need to build a quick-burning fire to take the chill out of a room, or to put a slow-burning log in the fireplace that will warm you all day, make the fire you build safe. Knowing that it is makes all your effort to find out what's best to burn worthwhile.

# Chapter 3

# Physical Conditioning and Body Use

Working with heavy logs and green wood teaches the woodcutter respect for their size and weight. The beginner, however, may approach woodcutting with more enthusiasm than common sense, so a few cautionary words are in order.

For anyone who wants to engage safely and enjoyably in the heaviest, most demanding woodcutting activities described in this book, there are two prerequisites: good physical condition and intelligent body use. Fortunately, any healthy man or woman can achieve the kind of physical condition needed and learn the skill of using the body sensibly and safely. Knowing that you are working well and in a way that prevents injuries to yourself and others makes woodcutting a totally enjoyable activity.

## The First Step toward Fitness

If you decide that you need to undertake physical conditioning, begin with a good medical checkup. Choose a doctor you can talk to—one who believes strongly in the value of exercise and activity and is interested in different kinds of physical conditioning. Tell the physician before the examination that you're planning to exercise strenuously in order to be able to work hard outdoors in cold weather, and describe the specific exercise program you want to follow. If the doctor has doubts about your readiness for the program you outline, ask him to recommend a program appropriate to your level of fitness.

# General Physical Conditioning

Two major types of physical conditioning are necessary for someone who wants to engage safely in the hardest and heaviest woodcutting activities: first, general respiratory, circulatory, and muscle-toning activity; second, activity that develops the ability to do hard physical work with the large muscle groups in the shoulders, chest, arms, back, abdomen, and legs.

Many people today are doing well indeed with the first type of physical conditioning. Running, bicycling, swimming, and vigorous walking are all excellent general body toners, especially when supplemented by one of the many fine indoor calisthenics programs available. Every public library can provide information on general physical conditioning and calisthenics.

# Work-conditioning for Woodcutting

However, the other kind of physical conditioning needed for safe, heavy woodcutting doesn't seem to be practiced or talked about much now. Perhaps this is because so much work today is sedentary, and makes little or no demand on the body's larger muscle groups. This kind of conditioning I will call "work-conditioning." If you want to enjoy a demanding outdoor activity like heavy woodcutting in your spare time, it is necessary to engage in work-conditioning deliberately unless you do hard outdoor work day in and day out, year-round.

Work-conditioning should certainly not be undertaken before a minimum of a month of general conditioning, and it does not replace general conditioning. The two go well together. With a little experimentation and a lot of motivation, it's not at all difficult to develop an individually tailored, combined program that doesn't take more than thirty minutes a day.

Use patience and common sense in establishing your pace. Regardless of what you may have achieved in the way of physical fitness and strength during some previous stage of your life, it is necessary to proceed gradually with work-conditioning. If you interrupt your exercise program, don't resume immediately at the level reached before the interruption. If the layoff has been for more than a week or two, go back to the beginning. Work forward at a pace that avoids soreness. Don't regard woodcutting itself as an interruption, though. The chief purpose of work-conditioning is to make it possible for you to move directly, and safely, into vigorous outdoor work which provides comparable *natural* exercise.

I, myself, became aware of the need for work-conditioning in my mid-thirties. I was working full time as a college teacher. I got my exercise walking across campus. I was doing no calisthenics. In my spare time, I worked on a 16-foot plywood camper I was building to go on a ¾-ton pickup truck. There were the usual interruptions that go with an indoor occupation. Sometimes weeks would pass before I could get back to extended work on my major project.

The result was my first bout of severe lower back pain. An orthopedic surgeon diagnosed it correctly as tendonitis, but gave me no long-range advice about what to do to prevent its recurrence—except for the discouraging general observation that although your muscles retain a good deal of *strength* as you get older, the ligaments and tendons inevitably lose elasticity. In itself, this is unquestionably true. But for people in their thirties, forties, and possibly many years beyond, losses of strength and elasticity due to aging should not be confused with those due to simple disuse.

Eventually, I discovered that proper exercise and body use can postpone indefinitely the need to curtail outdoor work. I am not as strong as when I was twenty. When I go camping with my teen-age son, I don't expect to have his resilience. But as long as I follow a program of general physical and work-conditioning, I can work as hard as I want at woodcutting—and that's good enough for me!

## Isometrics

Over a period of some fifteen years, I have found isometric exercises with a spring-loaded device called the Bullworker a convenient way of staying in condition for hard outdoor work. This device, and others like it, can be ordered from the Sears catalog. Like all apparatus, it takes quite a while to get the hang of using it most profitably, but it is light and compact, requires little room to work with, and comes with excellent instructions.

While some isometric exercises require the purchase of equipment or the construction of simple apparatus that can be made at home from materials on hand, most isometric exercises are designed to be done without any equipment at all, which makes them convenient for those who must do work-related travel. Your public library can acquaint you further with this form of exercise.

Isometric exercise is not a substitute for more general conditioning and has been disparaged by writers concerned exclusively with improving cardiovascular condition. Like any exercise that works the muscles hard, it has to be engaged in prudently, but my experience indicates that it can play a valuable role in maintaining strength and muscle tone for strenuous outdoor work.

# Training with Weights

For an adult who wants to engage in the full range of woodcutting activities, but who has not previously worked or exercised hard at strength-building activities, I would suggest training with weights as an alternative to working with an isometric system.

Weight training has long been used routinely for training in competitive sports. But until recently, most people did not have access to equipment. The recent surge of interest in fitness has made weight-training equipment widely available at Ys and health clubs. Both men and women can be found working out with barbells and the Universal Gym.

Work-conditioning with weights is not the same as body building for appearance's sake, though the same equipment and basic exercises are used. In body building, heavier weights are employed, with fewer repetitions, and workouts take much longer. Competitive weight lifting is still another matter, and has even less resemblance to work-conditioning with weights.

Training with weights as a phase of work-conditioning is quite simple:

1. *Always warm up thoroughly before exercising with weights and do some form of circulatory and respiratory exercise afterward.* One way of accomplishing this is to do the calisthenics part of the Royal Canadian Air Force program before using the weights, and the running part of the program after working with the weights.

2. *Do a complete set of basic body building exercises.* Though adequate instructions come with sets of weights, if you're working on your own it is a good idea to be familiar with a book like Jim Murray's excellent *Weight Lifting and Progressive Resistance Exercise.* The best way, of course, is to begin under competent supervision.

3. *Work with the lightest possible weights for a minimum of two weeks.* Strive for the best possible *form* in each exercise: good timing and balance, and full extension when

lifting or pulling. Breathe deeply, in a deliberately exaggerated manner, while exercising. Never hold your breath.

4. *Go through each exercise rapidly, with no long pauses between exercises.* All of the work with the weights should last no more than fifteen or twenty minutes.

5. *After the first two weeks, begin adding weight—but only enough to permit a minimum of fifteen repetitions for each exercise.* Two weeks later, increase the weight again to permit a minimum of ten repetitions in good form. Just enough effort should be required to leave you agreeably fatigued after the whole workout. When you are able to do a minimum of fifteen repetitions easily for an exercise with a given weight, increase the weight to the ten repetitions point.

6. *Work with the weights no more than three times a week.* After a long interruption, go back to the beginning and work forward at a rate that prevents soreness or discomfort. Never push yourself to the point of exhaustion.

# Coping with Muscle Soreness and Injury

Good physical condition helps to prevent injuries. Occasionally, however, there will be soreness from working or exercising improperly. Heavy exertion without an adequate warm-up, working or exercising in poor form, or simply not allowing for one full day of rest a week, are the most frequent causes of soreness and injuries.

After some experience, you will know whether it is advisable to continue on a reduced level of activity or rest completely when feeling sore. The most conservative course is to suspend all calisthenics, and certainly all forms of work-conditioning, until the soreness disappears. Rest, perhaps aided by heat, should be all that is needed. During the layoff, figure out what went wrong. When full exercising is resumed, increase the effort *gradually*. If problems arise again, get medical advice.

# The Reward of Discipline: Woodcutting in Comfort

I hope this program of general physical and work-conditioning doesn't sound too formidable. It needn't take much time—half an hour a day should be sufficient—and there are many physical and mental benefits. I have concluded that for every half-hour of conditioning I engage in when I can't "get away," I'm earning hours of woodcutting fun.

# Safe Body Use on the Site

The purpose of physical conditioning is to acquire the *strength* to work safely at woodcutting. Besides conditioning the body adequately, you must learn to use it properly while actually working. Underlying all my specific recommendations on body use while loading wood, cutting with the saw and axe, and splitting, are these general principles:

1. *Strenuous woodcutting should be engaged in only after an adequate period of complete conditioning.* Whether the conditioning is from formal exercise or comparable outdoor activity is immaterial as long as it includes both general conditioning and work-conditioning.

2. *Just as when exercising, always warm up before engaging in heavy or strenuous work.* If, when you're training with weights, you pick up a barbell without warming up, you're setting yourself up for a sore back. Exactly the same thing is true when gathering wood. When you see a nice load of wood left on the sidewalk for waste collection, don't park, pick the wood up immediately, and toss it in the back of your vehicle. Walk around for a few seconds. Rearrange things a

bit in your vehicle. Bend over from the waist a few times and move aside twigs and pebbles—even if none are really in the way. Even then, take your time moving the wood.

3. *Avoid excessive exertion.* Don't overwork when cutting wood. Know your own capabilities as a result of deliberate conditioning or experience with comparable work. "Feeling good" is usually a sign that one is not pushing the body too hard; the opposite should be looked upon as a warning of possible abuse. Although there is no single absolutely reliable indicator that will tell you how much is too much, your attitude can provide some clues. Watch out for feelings of negativity, panic, or of anger at the wood. If these occur, stop work until they go away.

4. *Prepare by thorough conditioning before working, but when actually cutting wood, do everything in the easiest way you can.* Traditional woodcutting tools have been perfected in design and operation during centuries of use by people just as clever as we are. Such tools are "clumsy" or "crude" only for those who are in inadequate physical condition to be doing full-scale woodcutting anyway, or who don't know the right way to use them.

All my work to date supports this finding: when a traditional hand tool is difficult for me to work with, the fault is mine, not the tool's. I'm not in condition to handle it properly. I don't know how to repair, sharpen, or adjust it. I'm not clear about how to use it. When these faults are corrected, the tool becomes a pleasure to use, and I find myself marveling at the cleverness of its long dead, usually anonymous, designers.

This can be only the beginning of a thorough discussion of physical conditioning and body use. There is any amount of excellent reading on physical conditioning that is easy to come by. When it comes to the use of the body in intrinsically satisfying outdoor work, I can think of much less.

There is, however, a very old Chinese philosophical fable that suggests enough to go on for a lifetime. The hero is a meatcutter, not a woodcutter. I always think of it when cutting wood—or doing anything else with hand tools. It's been in my head for years now, and I still believe that if I could only get to the bottom of it, nothing else would have to be said about the best way to work. Here it is. See whether you agree!

A man who was cook to Prince Wen Hui was cutting up a bullock. The blow by his hand, the thrust from his shoulder, the stamp of his foot, the heave with his knee, the whish of the flesh coming away, and the whistle of his knife going in were all perfectly in time, having the rhythm of the Dance of the Mulberry Grove, and exactly in time like the chords of the *Ching Shou.* The Prince said, "How excellent, that you should reach this pitch of perfection!" The cook put down his chopper and said, "The thing that your humble servant delights in is something higher than art, namely the Tao. When I first began to cut up bullocks, what I saw was just a bullock. After three years I no longer saw a bullock as a whole, and now I work by the spirit and not by the sight of my eyes. My senses have learnt to stop and let the spirit carry on. I rely on the Heaven-given structure of a bullock. I press the big tendons apart and follow along the big openings, conforming to the lines which must be followed.

"A good cook gets a new chopper every year; and this is because he cuts. A poor cook gets a new chopper every month; and this is because he hacks. But the chopper belonging to your humble servant has been in use for nineteen years. It has cut up several thousand bullocks, but its edge is as keen as if it had just come from the whetstone. There is a space between the joints, and the edge of the chopper is very thin. I put this thinness into the space, enlarging it as I go; and there is bound to be plenty of room for the blade.

"Nevertheless, every time I come to something intricate, I take a look at the difficulty. Apprehension calls for caution. My eyes dwell on it, and I act very slowly. The movement of the chopper is imperceptible, and by degrees it all comes away like bundling soil from the

earth. I then lift the chopper out and stand up and look
all around with the satisfaction of victory. Then I wipe
the chopper and put it away."

—Chuang Tzu, in E. R. Hughes,
*Chinese Philosophy in Classical Times*

# Chapter 4

# What to Wear

The body is the innermost part of *the material self* in each of us; and certain parts of the body seem more intimately ours than the rest. The clothes come next. The old saying that the human person is composed of three parts—soul, body and clothes—is more than a joke.

—William James, *Principles of Psychology*

William James is right; clothing is no joke, particularly when you're dressing for heavy woodcutting away from home in cold weather. It can still be part of the fun, and all the usual elements of personal choice can come into play. In selecting your garments, however, make sure that you consider first safety, suitability for hard work, and protection against the weather.

In every kind of woodcutting it's important to give thought in advance to how you're going to protect yourself from injury. Physical fitness is as important as intelligent body use in protecting yourself from muscle strains or sprains. Clearing the area where you will be woodcutting of broken branches and loose stones will help to prevent a fall. Wearing sturdy footgear and gauntlets or gloves on your hands is another way to help prevent accidents.

# Protecting Your Hands

Wearing gauntlets—heavy leather gloves with wrist protectors—will help to ensure a safe grip as you swing the axe down to limb a tree, or load logs into your vehicle. In addition, wearing gauntlets that fit well can help to prevent your hands from blistering as they move on the axe handle with each stroke.

Gauntlets can be purchased in hardware stores and lumberyards. They are expensive, so take care of them. If they get soaked, dry them slowly, not close to strong heat. When you're wearing gauntlets, brush off particles of sand or dirt before tackling heavy wood. If you neglect to clean off the wood you handle, you can wear out your gauntlets in the fingers—even though the rest of the glove is still good.

If the stitching goes, but the leather and cloth parts are sound, a gauntlet can be repaired by a good shoemaker. If you would like to do your own repairs, inexpensive supplies and instructions for resewing leather can be obtained from a leathercraft shop.

Wool inner gloves or mittens should not be used for woodcutting without leather outer mittens, which protect them from wear. There is no difficulty in working in weather 20 degrees below zero with such a combination, provided the fit is not too tight and the wool liners are kept clean and dry. This may mean having to switch to a pair of good leather work gloves as your hands begin to perspire.

Leather work gloves are useful in any weather, while working close to home as well as in the woods. They, too, are best purchased from a place that caters to tradespeople.

Like the gauntlets, good work gloves aren't cheap, so keep track of them in the woods.

# Protecting Your Feet

Protecting your feet is just as important as taking care of your hands. There is one fundamental rule: *never cut or handle wood without sturdy shoes or boots on your feet.* To wear sneakers or light footwear is to invite painful, unnecessary injury.

The best protection when loading and unloading logs is that provided by steel-toed work shoes. With caution, you can make do with good quality work shoes or boots that don't have steel toe inserts. There's no need to worry about the type of sole. Good quality work shoes and boots—the only ones worth buying, of course—come with adequate soles and heels. Given the cost of hiking boots and the likelihood of scratching or abrading them while cutting wood, I wouldn't wear hiking boots for this kind of work.

Purchase work boots or shoes as you would hiking boots, using the same guide for size. Allow for thick socks and a heavy load on your back when trying them on. I prefer to use two pairs of socks—a thick wool boot sock with a cotton work sock worn underneath. Keep in mind that in cold weather your toes must be able to move freely.

For leather work shoes or boots used in snow, I like an old-fashioned method of waterproofing: plenty of neat's-foot oil to waterproof the leather uppers and keep them pliable, and dubbing to seal the seam between uppers and soles. Neat's-foot oil, dubbing, and the newer types of waterproofing materials can be obtained from a shoemaker or leather repair shop that handles riding equipment.

There are some interesting old recipes for waterproofing lotions and unguents in Horace Kephart's *Camping and Woodcraft*—but the materials he calls for are often hard to find. I can get the materials for the recipe that follows, but I doubt whether my family would like to have me test it in the kitchen:

> Boil together two parts of pine tar and three parts of cod liver oil. Soak the leather in the hot mixture, rubbing in while hot. It will make boots waterproof, and will keep them soft for months, in spite of repeated wettings. This is a famous Norwegian recipe.

# Selecting Clothing for Comfort and Durability

After providing protection for your hands and feet, make sure that your other clothing is suitable for hard outdoor work. Outer garments should be tightly woven, but not tight fitting. The stitching should be strong, and preferably with sufficient seam allowance for resewing when necessary. The fabric of your outer garments should not catch readily on twigs or brush—either because of texture or bulk—and should, of course, be wear resistant.

You need plenty of room for body and arm movement, as well as for other clothing underneath, but there should be nothing dangling. Leave jewelry and tie belts at home.

No matter how cold the weather, you will warm up fast once the actual work begins. Dress so that outer garments can be removed, a layer at a time, to prevent heavy sweating. The colder the weather, the more important it is to strip as you heat up. The principle is the same as when hiking or snowshoeing in cold weather; peel off your outer garments as you warm up, replace them when resting.

# Color Notes

For your own protection, dress for visibility during the hunting season, when people with firearms may be moving through wooded areas. Should this be a possibility where you are working, wear something red all the time. If you're carrying a pack, have a good bit of bright color on it, too. Mark your tools as well. Even when there is no danger of being mistaken for a deer, a bit of bright color on clothing and equipment used in the woods makes it harder to mislay things.

# Chapter 5

# Tools and Equipment

Let the reader beware! I write even less objectively about tools for woodcutting than about physical conditioning. I have an ideal of tool use, and I'm going to propose it fervently—but it wasn't always my credo. A full account of my wallowings in the sloughs of acquisitiveness and mechanization would have some of the complexity, though none of the importance, of St. Augustine's *Confessions.*

For example, I now urge holding off when it comes to purchasing a chain saw, or even better, doing without it altogether, but that's not what I did myself. Early in my woodcutting career I bought a chain saw and used it intensively before I became aware of its dangers and came to prefer hand tools.

And though I am given to warning solemnly against "collecting" tools, as opposed to buying only what one needs at the time, there was a period of no little duration when I would practice any kind of sophistry upon myself to persuade myself that I *needed* something I saw in this or that favorite store or catalog.

It is not that I don't want to be taken seriously now when I say that the ideal ought to be to use as few tools as possible, but I must admit that I came to this conclusion by making my own mistakes. This view has brought me increasing satisfaction, and I expect to stay with it the rest of my woodcutting days.

This chapter will present an overview of what can be used by the amateur woodcutter. I won't mention anything I haven't used myself, but that doesn't mean that someone else needs all the tools and equipment I describe, particularly when just starting out. Since the major emphasis in the book is on the advantages and satisfactions of working with hand tools, I won't include at this point a discussion of chain saws and their accessories; there is a separate chapter on this subject further on.

# Cutting and Splitting Tools

The *cutting tools* are axes and saws. The world of axes is large and fascinating, but we can keep matters simple. The basic *length* is the full-size handle, about 30 inches long. The basic *weight* is a 3½-pound head. The best *general type* is a single-bitted axe, which has only one cutting edge. Conventional wisdom—with which I concur—is that for safety's sake a beginner should not work with a double-bitted, or two-edged, axe. It is too easy to stumble and cut yourself on a razor-sharp edge.

Years ago, a great variety of shapes was available among 3½-pound, single-bitted axes. There isn't as much choice any more, but the beginner doesn't have to worry. The important thing is to work with a good axe that is properly sharpened and correctly fitted to its handle. For more information about this, see the section on replacing an axe handle in Chapter 6, "Caring for Tools and Equipment."

Campers are familiar with the somewhat smaller three-quarter axe. Besides having a shorter handle, it has a lighter head, usually no more than 2 pounds. It has its uses in wood-cutting, but does not replace the full-size axe if you need to cut and split a lot of fair-size wood since, with its lighter head, you have to do more physical work. At least one first-rate outdoor writer believes that the three-quarter axe is too dangerous for a beginner to use for cutting up logs:

> The three-quarter axe . . . is the cause of many accidents in the woods, even among experienced woodsmen. Owing to the shorter length of the handle, the axe may not swing clear of the body after a miss, and serious injury to an ankle or foot may result. A missed stroke with a full-handled axe will either swing clear of the body or end in the chopping block . . . It is not a tool for the beginner.
>
> —Rutstrum, *The New Way of the Wilderness*

In the same family as the axe is the hatchet. It's even less likely than the three-quarter axe to be used as a primary woodcutting tool, but I consider several of the hatchets I own useful accessory tools. The light boy scout type, if it is of good quality, is excellent for cutting up smaller branches when working in the woods. A heavier hatchet—the older the better, it seems—is useful for splitting wood in the back-yard.

Hand saws used in woodcutting are considerably differ-ent from those employed in carpentry and cabinetmaking. The two general types of hand woodcutting saws are the crosscut saw and the bucksaw. The woodcutting crosscut saw comes in two sizes. The large version has to be worked by two people. The smaller crosscut can be employed by one person alone or, with the addition of an extra handle that comes with it, by two people.

Woodcutting crosscut saws are used for cutting up large trunks and good-size logs; some people prefer them for felling trees (for more on crosscut saws, see Chapter 9, "Bucking with the Log Saw").

The other general type of hand woodcutting saw, the bucksaw, is used for cutting wood to burning length. The wood is supported in a homemade wooden brace called a sawbuck. The bucksaw is a fine, traditional tool and a pleasure to work with; it should not be confused with the flimsy modern tool called the "bow saw" (there is a full discussion of the bucksaw in Chapter 11, "Cutting Wood to Stove or Fireplace Length").

Splitting tools include a sledgehammer and a set of steel wedges. This combination is required for the most difficult situations. Working with easier wood, you can get by with one wedge and a splitting maul. The splitting maul has a head with one end shaped like the cutting edge of a wedge, the other like a sledgehammer. Axes, either with the 3½-pound head or with 4½- to 5-pound heads, can also be used for splitting (Chapter 12, "Quartering and Splitting," has the details).

For cutting and splitting tools to be safe and efficient, they have to be properly sharpened. A large, hand- or foot-operated grindstone is best for shaping edges and removing nicks in axe blades. The next best arrangement is a set of metal cutting files, and a sturdy metalworking vise for holding the axe while filing. Least desirable is a power-driven grinding wheel, particularly the small, high-speed type.

When the edge has been ground or filed to shape, it is then honed with a Carborundum stone lubricated with oil. When working in the woods with an axe, you should carry a small mill file and a small, circular Carborundum stone.

Saws are also kept sharp with files. Complete resharpening requires a few simple accessories that can be made at home.

Keeping your cutting tools sharp while working with them, and resharpening them completely when necessary, is an essential part of learning to use them. Chapter 6, "Caring for Tools and Equipment," will get you started.

# Equipment for Moving and Loading Wood

The simplicity of the chain, rope, block and tackle, and wheelbarrow should not mask their immense value for the amateur woodcutter. Detailed information on the use of such equipment is in Chapter 7, "Transporting and Storing Firewood."

# Vehicle and Vehicle Accessories

When you move farther away from home, particularly in cold weather and rough terrain, it is wise to have enough vehicle equipment to deal with immobilizing situations. The best protection, of course, is knowing how to avoid them altogether; but that takes experience, and in the meantime you want to be able to cope with problems that needn't be serious if you are prepared.

A four-wheel-drive vehicle with front-mounted winch may be the best solution if you've got the money, and you know how to use such a rig in the field. But make your decision thoughtfully. Safely winching a badly stuck four-wheel-drive vehicle out of deep mud or snow requires more special know-how than is needed to keep a humbler vehicle from getting bogged down in the first place—and about as much accessory equipment, too. With an ordinary vehicle you *know* from the start that you have to be careful all the time in rough country, especially in the winter, and *that* attitude is your best protection.

No matter what you are driving, the following should be considered *minimum* vehicle equipment once you leave the main roads:

- adequate tire-changing equipment, including good spares for vehicle *and* trailer; a heavy-duty (at least

3-ton capacity) hydraulic jack; sound pieces of
timber to use under the jack and as blocking for
wheels; and a good wrench for changing tires
- digging tools: pick and shovel
- strong tow chain with hooks or strong tow cable
- 6-volt lantern with fresh battery

In remote areas, you won't go wrong if you also carry:
- heavy-duty, outside-mounted gas can filled with
  extra gasoline
- funnel for transferring gas from can to vehicle tank
- tire chains
- basic vehicle repair tools
- basic spare parts, including distributor cap, rotor,
  spark plugs, high- and low-voltage wire, fan belt,
  headlight, flasher, fuses, and bulbs
- sleeping bag
- canteen with water
- food
- matches

If this sounds like an outfit for a wilderness expedition,
it's just as well to be impressed forcefully with the principle
that the farther you remove yourself from immediate help,
the more self-reliant you should be. The tire on a trailer *can*
go flat; a heavily loaded car or van *can* get stuck in mud or
snow; water *can* splash up into your distributor and crack
the cap. Having the equipment to deal with such problems
—and having practiced in advance how to use it—will only
make you that much more likely to avoid such difficulties.
But if something does happen late on a Sunday afternoon
when you're miles away from help, it's good to know that
you can handle the situation. The alternative can be worry-
ing to yourself and the folks back home, time-consuming,
and expensive.

Adequate tire-changing equipment is very important.
The jack furnished with a new vehicle is usually too flimsy
to be used safely in the field. A light bumper or scissors jack
can be awkward enough on level pavement, but it's danger-

ous to use a light jack with a loaded vehicle on rough, uneven ground.

As for the factory-furnished tire wrench, even a pretty strong person may find it hard to work with. Cold weather makes matters worse. I always carry an extra swivel-headed socket wrench and an extension handle. With that combination, anyone can deal easily with the tightest wheel nut. Almost as good—and cheaper, and easier to obtain—is a large four-way lug wrench, one with 20-inch arms.

The safest method of changing a tire in the field is to work with a good hydraulic jack and proper blocking. Practice using the jack correctly before an emergency occurs, but do it under the guidance of someone who knows how to work with a loaded vehicle on uneven ground.

A good tow chain is nice to have even if you never get stuck. Years ago, I bought a fine one, complete with hooks, from a blacksmith in Alberta. Despite a fair amount of traveling in the bush with a heavily loaded two-wheel-drive pick-up truck, I never had to use it on my own vehicle. But I've helped other people more than once, and that's a satisfaction, too. It's just as important to be able to help others as yourself when you get away from "civilization."

# Personal Gear

Before you begin moving into the woods, you should consider how to transport your cutting tools safely and what equipment you should carry. Don't worry—despite my heavy advice about vehicle equipment, I'm not going to urge that you go 100 yards into the woods equipped for a week in a wilderness area—but a little of the knowledge and much of the spirit of backpacking is in order. A pack frame with a load-bearing shelf is very useful. Whether you are carrying a load of wood or not, you can lash a poncho, axe, and rope or chain onto the frame. With these items secured on the pack frame, you will have your hands free to carry a crosscut saw. Instead of thinking about the awkward bundles in your arms, you will be able to watch your footing and concentrate on safety.

Though I wouldn't argue that it's absolutely necessary, I see no harm in abiding by the old woodsman's rule of always having in your pockets or on your back a first aid kit, a pocket sewing kit, matches, and a knife. And don't forget a big red bandanna, an item of a hundred uses in the woods.

Expert opinion has been shifting lately on the importance of carrying a snakebite kit in warmer weather. The best advice I can offer is to follow the recommendations of local physicians and wildlife experts.

I carry a pocket snakebite kit. But in more than thirty-five years of hiking, camping, and woodcutting, I have never been bothered by a poisonous snake. Long ago, however, I was taught where to put my hands and feet, and when silence is *not* a virtue. The old rule is still valid: leave them alone, and they'll do the same to you.

Nonpoisonous snakes should be left alone for several reasons. They have as much right to be there as we have, and are usually a lot more helpful than we are in the ecology of forests and fields. To attempt to kill a poisonous snake is looking for trouble; to kill a harmless snake is a demonstration of hysteria or ignorance. Learn to tell the good guys from the bad guys, then leave both kinds alone.

# Deciding What Tools You Want to Use

I have surveyed the tools and equipment most likely to be used in hand woodcutting. Now the important question is not what might be used, but what is personally necessary. To answer the question, consider first where you will be doing your woodcutting. Foraging for wood in your immediate neighborhood or no farther than the city limits calls for one type of woodcutting outfit. Driving into the woods alone in midwinter, miles from home, requires a different kit.

How do you want to do the cutting? You can make the chain saw central in your cutting operations, or decide to use only hand woodcutting tools. To opt for the chain saw does not mean that the complete outfit will contain fewer tools and less equipment; it certainly doesn't mean that you will spend less!

If you decide to work with hand woodcutting tools, you must choose which tools you want to use. You may wish to use a single axe as your only cutting tool, and perhaps as the only splitting tool, as well. Acquiring the necessary skill will take more time and patience than any other approach, but it can be done. Most hand woodcutters will use a bucksaw, a pair of wedges, and a sledgehammer or splitting maul. If you include a crosscut saw in your tool selections, you make yourself completely independent of the chain saw.

# Acquiring Tools—by Retail Sales or Salvage

Finally, when you have decided what tools and equipment you need, you can think about acquiring them. Today's prices being what they are, I hope you don't have to buy much. When you do make purchases, ignore novelties and miracle claims. Buy *only* from dealers who can attest to the quality of what they're selling, or from reliable mail-order firms, like the ones listed here. Well-made tools and equipment properly cared for will last more than a lifetime.

The Ben Meadows Company
3589 Broad Street
Atlanta, GA 30341
(404) 455-0907

Ben Meadows *General Catalog* of "Forestry & Engineering Supplies" offers a good selection of hand tools for the woodcutter: bow saws, one- and two-man crosscut saws, a good selection of Sager and Hudson axes, sheaths, replacement handles, sharpening stones, wedges, and sledgehammers. They also sell chain saws.

Garrett Wade Company
302 Fifth Avenue
New York, NY 10001
(212) 695-3360

The Garrett Wade catalog provides well-written descriptions of the tools available, accompanied by photographs that clearly show their quality. Garrett Wade stocks a wide selection of domestic and imported hand and power tools for the fine woodworker and carpenter. For the woodcutter, they have one cross-cutting log saw, a saw set, sharpening stones, and saw files.

Glen-Bel Enterprises
Route 5
Crossville, TN 38555

For $3, Glen-Bel will send you their *Country Store Catalog,* listing "over 3,000 items of hard-to-find brand new general merchandise." These items include woodlot tools such as axes, wedges, mauls, sledgehammers, bow saws, one-man crosscut saws, logging chains, winches, and pulleys.

Lee Valley Tools, Ltd.
P.O. Box 6295
Ottawa, Ontario
Canada K2A 1T4
(613) 728-4625

Lee Valley offers a selection of the best quality tools in the world for the serious woodcarver, cabinetmaker, and logbuilder. The woodcutter can order a crosscut log saw, bow saws, sharpening stones, replacement handles, and a wide variety of vises and clamps for the workbench. Those interested in the history of woodcutting will find the reprints of old tool catalogs exciting. For example, the *Henry Disston and Son 1914 Catalogue* illustrates ". . . the 38 different two-man crosscuts they then made, complete instructions for setting and sharpening cross-cuts . . ."

Nasco
901 Janesville Avenue
Fort Atkinson, WI 53538
(414) 563-2446
Free Phone Order Service
Dial 1-800-558-9595

Although Nasco is primarily interested in supplying the dairy farmer and cattle rancher with equipment for care of his herds, they do have some forestry equipment available, including a few bow saws and axes, a number of chain saws, and steel, aluminum, and plastic wedges.

Northeast Carry Trading Company
110 Water Street
P.O. Box 187
Hallowell, ME 04347
(207) 623-1667

Northeast Carry Trading Company is a membership organization ($6 fee) providing a catalog of quality tools and hardware, a lending library, and workshops and consultation for the small landholder interested in greater self-sufficiency. Although they do not have a large selection of hand woodcutting tools, you can mail order a bucksaw from them.

Woodcraft Supply Corporation
313 Montvale Avenue
Woburn, MA 01888

Besides offering a large selection of woodworking tools, Woodcraft offers a number of hand woodcutting tools. Their selection of saws includes the bucksaw and one- and two-man crosscut saws. Woodchopper's mauls and splitting wedges are also available.

Buying tools secondhand or at flea markets is not a good idea unless you know exactly how the tool is used and what needs to be done to put it in first-class working order. You should also know how much a tool of comparable quality costs new. Don't be misled by uninformed chatter about what isn't being made anymore; find out for yourself by studying the catalogs of the better dealers.

There are a lot of old, discarded tools around that can be put back into use. Rust doesn't help an axe's appearance, but it takes an enormous amount of neglect to render the tool completely useless. New handles can be fitted to axes and saws. Unless there is actual breakage or the teeth have been resharpened far too many times, a lot can be done to make saws that haven't touched wood for half a century usable again. Mushroomed wedges can be rehabilitated (see Figure 6-8 in the next chapter). Rope that has not been overused or stored damp remains sturdy a long time. An old block and tackle can be just as good as a new set, and you don't need a country blacksmith to fit yourself up with a logging chain.

Fortunately, there are people who can be found who do wonderful things with forging, brazing, welding, grinding, and retempering—if they like you and the tool well enough. But for the woodcutter who has chosen to work with traditional tools, learning how to restore these older tools for your own use is an immensely satisfying hobby. The next chapter, "Caring for Tools and Equipment," provides enough information to get you started—and keep you going—for years.

# Chapter 6

# Caring for Tools and Equipment

Tzu-kung asked how to become Good. The Master said, a Craftsman, if he means to do good work, must first sharpen his tools . . .

—Confucius, *Analects* XV:9

Keeping your hand tools and simple equipment in good condition is part of the fun of working with them, but things being what they are today, it's also a necessity for most of us. Perhaps in a rural area you can find someone who for a fee will "whip" the ends of lengths of rope, resharpen a new axe correctly for serious cutting, and carefully touch up the teeth of a saw that is just beginning to cut a little hard. But in the average city, town, or suburb, you'll spend more time trying to locate such a person than doing the work yourself.

The best time to begin proper tool care is when you acquire the tool. Start by doing the simple but necessary things right away; your work will be safer and more pleasant, and you'll provide yourself with a basis for more advanced maintenance when it becomes necessary.

# Manila Rope

Good quality Manila rope can be purchased where tools and contractors' supplies are sold, and also from marine hardware dealers. It is superior to artificial fiber ropes for woodcutting purposes, provided a few simple principles of caring for it are observed.

The time to begin taking care of your rope is when you buy it. Keep it neatly coiled, clean, and away from objects which could cut it.

The dealer will probably tape the ends to keep them from unraveling. When you get home, decide on the lengths you need and divide up the rope with a sharp knife. The next step is to "whip" the ends of the pieces you'll be working with. Whipping means wrapping heavy thread around the ends of the rope; this holds the strands of the rope together permanently. Of the many ways of whipping, Figure 6-1 shows one of the simplest.

When you're not using your rope, store it in a dry place, loosely but neatly coiled, and with nothing heavy on top of it. Manila rope is not bothered by heat, but since it is a natural fiber, it will rot if stored damp.

After you have used rope in the field, shake it clean of dirt and sand, which act as abrasives on the fibers. It's a good idea to wash the rope in clear water every now and then, and let it dry in the sun.

Using proper knots is also important for keeping your rope in good condition. Acquaint yourself with a few basic knots such as those shown in Figure 6-2.

Using proper knots is also important for keeping your rope in good condition. Acquaint yourself with a few basic knots such as those shown in Figure 6-2. With a bit of practice, you can easily learn to tie and untie these knots quickly. There are, of course, many other knots; only a few have to be added to those shown to provide a lifetime repertoire for woodcutting.

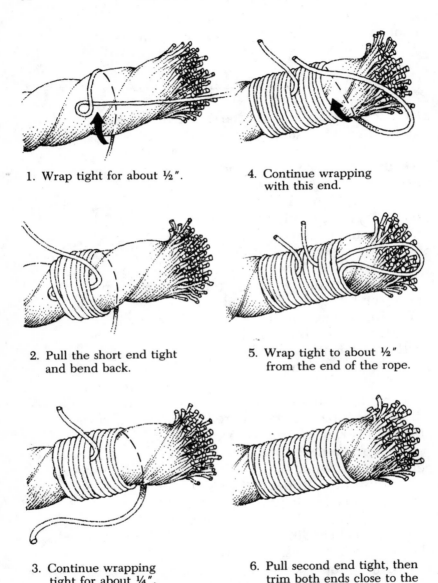

1. Wrap tight for about ½".

2. Pull the short end tight and bend back.

3. Continue wrapping tight for about ¼".

4. Continue wrapping with this end.

5. Wrap tight to about ½" from the end of the rope.

6. Pull second end tight, then trim both ends close to the wrapping.

**Figure 6-1:**  Use any good twine to whip rope. When actually whipping the rope, keep the turns as close together as possible. Proceed in steps described above.

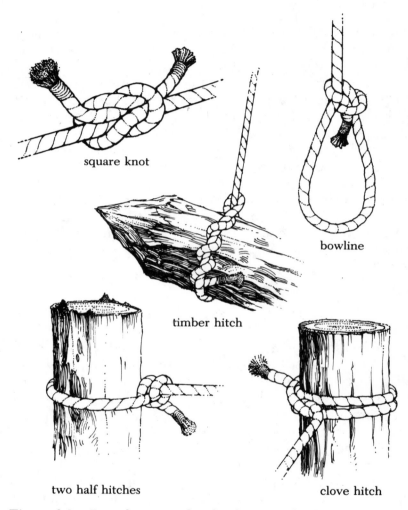

square knot

bowline

timber hitch

two half hitches

clove hitch

**Figure 6-2:** Basic knots used in hauling wood off the site.

For more information about knots, their uses, and how to make them, look at Raoul Graumont's fine little *Handbook of Knots.* Besides taking you deep into this important traditional lore, this volume provides useful information on rigging blocks and tackles. The illustrations show how to make every imaginable knot, tie, and splice, and there is an

interesting "Index of Knots by Job and Hobby." Just glancing at the lists of knots used by forest rangers, hoisters, log-
gers, and lumberjacks is exciting; it shows how much can be
done with such a simple thing as rope.

There are, of course, good reasons why different knots
have been developed. A knot appropriate for a given purpose is safe and efficient. It is also easily tied and untied, and
helps preserve the working life of the rope. Never store rope
with any tight knots in it. Doing so will create a permanent
set in the rope, which weakens the fibers.

Working carefully with rope is as important as storing
it properly if you want to make it last. Be continuously alert
for sharp rocks and corners of any kind. When dragging a log
section, keep the rope off the ground by raising the tied end
slightly. If this isn't possible, use a logging chain instead of
a rope.

# Chain

Even though chain is made of steel and will wear like
iron, don't expect it to be able to endure anything. Do not
drag anything wrapped with chain over a paved road. Towing a loose chain through dry dirt or sand is a good way of
cleaning it; but *don't* do what I did early in my "logging"
career. I got the not-so-bright idea of towing a log circled by
my logging chain along a hardtop road. Something made me
stop my truck after a few hundred yards and go back to
inspect my ingenious labor-saving arrangement. I don't remember how badly I scratched the road, but I can still find
the links on my chain that were worn flat by this dumb stunt.

Chain may not have been much a part of your life before, but if you become an advanced amateur woodcutter,
you will probably find yourself working with tire chains,
logging chains, and tow chains.

Tire chains are wonderfully effective on snow and ice
when they fit correctly and are properly attached. It's a bit
of work to install them, though, and they will break if you

drive at speeds over about 25 miles per hour—two reasons why they have passed out of favor. But a set of rear wheel chains is a great comfort on ice or packed snow, where no one should be driving fast anyway; and the greatly increased traction they provide is well worth a few minutes of chilly communion with the ground while installing them.

Tire chains will rarely break if used sensibly, but if they do, they can be repaired with special repair links. There's even a little hand tool available for installing the repair links.

Strictly speaking, logging chains are for people who use draft animals or tractors to drag logs. But a light chain, say with 1¼-inch links, 12 feet long, with a grab hook at one end and a grab or round hook at the other end, is better than rope for pulling heavier logs along the ground, even if you do it by hand. Unless you badly misuse such a chain, it will never break, but it's worth knowing that a broken chain of this sort can be repaired, too, with repair links. Such links are available in the kind of hardware store where the proprietor knows the difference between a cutter mattock and a pick mattock.

In such a store, chain repair links will be only part of a larger collection of goodies classified as "chain tackle." Here, you'll find not only extra grab hooks and round hooks, but swivels and clevises. The latter are U-shaped fittings used to attach hooks or other special items to the ends of a chain.

A good tow chain is so heavy and strong that you're more likely to bruise yourself with it than to damage the chain by anything you do with it. It will be easier to handle and will stay cleaner if you keep it in a sack.

Should you lend your vehicle to anyone, be sure the borrower knows the difference between a logging chain and a tow chain. The light logging chain will snap if used for pulling a vehicle out of a ditch.

Speaking of snapping, whenever great tension is put on any rope, chain, or cable, keep yourself and everyone else well out of the way. Steel cable, which is also called "wire rope," is particularly dangerous in this regard, since a broken end can whip around viciously. Workers and bystanders should take into account where the vehicle or object being pulled may go if the rope, chain, or cable snaps suddenly.

As usual, the forest service's excellent safety manual is worth heeding in this regard:

Avoid sudden shocks and overloading.

Chains shall not be kinked or knotted.

After hitching or hooking chains or cables to such objects as logs, stumps, or machines, everybody shall stand clear and as far away from the tractor or load as the length of chain between them.

—*Forest Service Health and Safety Code*

(This is just a selection from the material on chain. The whole of this little book should be studied by anyone who plans to spend a lot of time working in the woods.)

# Recovering Knowledge of How to Sharpen Hand Tools

As our society has moved away from using hand tools, the knowledge of how to sharpen them has faded, too. This makes for some slight initial difficulty for those of us who see advantages in returning to the use of hand tools.

No two people work with a hand tool the same way, nor do they have to; the "simplicity" of the tool allows for infinite variation in personal style. By the same token, the more familiar you become with the possibilities of a hand tool, the more likely you are to develop personal techniques and preferences for sharpening it. This may be another reason knowledge of how to sharpen hand tools is dying out.

Sharpening isn't a chore once you get into it. It's an opportunity to get to know the tool intimately and, after a while, a chance to shape the metal of the tool just as you want it to be. Besides being interesting and enjoyable in itself, hand sharpening increases the pleasure of working with the tool. A well-sharpened tool is agreeable to work with; indeed, you could use pleasure in handling as a test of sharpness. Only too often today we confuse the laboring of a dull hand tool with the supposed difficulty of doing the work by hand. Sharpen the tool properly, and the improvement in ease of handling can be amazing.

I think a beginner should start right away with simple and easy sharpening tasks. Don't worry about mistakes; you cannot ruin the tool if you use only hand sharpening devices, and use them in moderation. Slight errors won't matter, and will probably be corrected the next time you work on the tool.

Even if you have never sharpened a tool, you can begin to teach yourself how to do it well with axes, wedges, and crosscut saws. These tools have large cutting edges or large teeth; this means you can see easily what you're supposed to do. Also axes, wedges, and saws are not tempered as hard as chisels and knives; the softer steel in them is not difficult to remove by hand filing.

# Equipping Yourself to Sharpen Hand Tools

Well then, how do you get started? Very little in the way of equipment is needed to sharpen an axe, a pair of wedges, or a crosscut saw, as you can see from the following list:

- an ordinary bench-type (1 × 7 × 2 inch) combination sharpening stone
- a pint of mineral oil (available from a drugstore)
- these files, as shown in Figure 6-3:
  - 12-inch double-cut bastard mill file
  - 8-inch single-cut smooth mill file
  - 8-inch single-cut smooth crosscut file
  - 7-inch single-cut smooth taper file

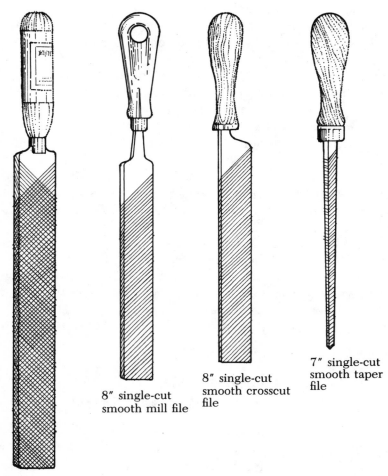

8″ single-cut
smooth mill file

8″ single-cut
smooth crosscut
file

7″ single-cut
smooth taper
file

12″ double-cut
bastard mill file

**Figure 6-3:** Files used in sharpening axes, crosscut saws, and bucksaws.

Although various field expedients are possible, a beginner needs to have the tool that is being sharpened securely clamped. At the start, arrange to have access to a bench vise, and a woodworking vise. Even the small, 3½-inch wide-jaw bench vise can be used, as long as it is solidly mounted to a sturdy bench. This vise is for holding axes and wedges while filing them and sharpening with the stone. The woodworking vise is for holding the crosscut saw in place while touching up and resharpening the blade.

Notice that I have not mentioned a power grinder. A high-speed grinder can easily ruin an axe (a chisel or plane iron, too, for that matter). If you want to use an electric grinder for axes and wedges, a more suitable one can be made from an old refrigerator motor (see Figure 6-4). The advantage of such a grinder, apart from its cheapness, is the

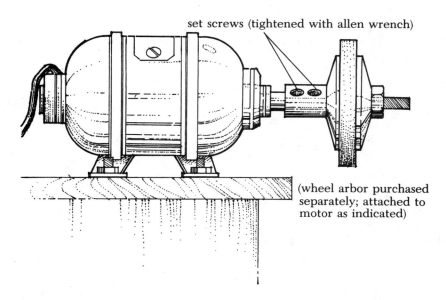

set screws (tightened with allen wrench)

(wheel arbor purchased separately; attached to motor as indicated)

**Figure 6-4:** To safely, slowly sharpen cutting tools, use a grinding wheel made from a refrigerator motor.

lower speed at which the motor runs. It takes longer to shape an edge, but there's less danger of destroying the temper. Even with such a grinder, however, the edge must be kept cool by dipping it in water frequently and keeping it in motion across the edge of the wheel. And for your own protection while grinding, build a plywood guard around the grinding wheel and wear a clear plastic face shield. For details on building such a guard and more information on using a homemade grinder, see Harry Walton's *Home and Workshop Guide to Sharpening.*

You don't *need* a power grinder for sharpening hand woodcutting tools. The right file, a solidly mounted vise in which to hold the tool you're working on, and an idea of what hand filing can do will enable you to dispense with any power grinding device.

Like so many other fine hand tools, files have been over-shadowed by power tools, so let me say a few words generally about them, before discussing their uses in sharpening tools.

The specification, "12-inch double-cut bastard mill file," means a file that
- is 12 inches long exclusive of the tang (the part that goes into the file handle)
- has its teeth cut in two directions, instead of one
- has its teeth spaced according to the distance conventionally called "bastard" (other spacing designations are "smooth" and "middle"—but this is not a complete listing)
- has the rectangular cross section and overall shape traditionally called "mill" (files of this shape were used to sharpen the teeth on large circular saw blades in sawmills)

Thus, the complete specification of a file gives its length without the tang; an indication of whether its teeth are cut from one or two directions; an indication of the spacing of the teeth; and a description of its shape.

We speak with good reason of the *teeth* of a file, since

it is a cutting tool, not a sanding or abrading tool. It takes a long time to wear out the teeth of a good file if it is properly used, but like any cutting tool it can be quickly dulled if handled carelessly.

A file is designed to cut only on the forward stroke, and should be lifted on the return stroke; if it is dragged back across the work, the teeth of the file become dull very quickly. Enough pressure has to be applied on the forward stroke to make the file cut the metal; you can tell whether the file is biting by the feel and the tiny chips produced. Great hand or body strength is not required to manage a file correctly. Even when filing a large piece of metal, like an axe or a wedge, with a large file, anyone can apply the pressure needed by leaning into the stroke with the weight of the body.

Two basic strokes are used in filing, as shown in Figure 6-5. First, there is the heavy forward cutting stroke just described. This is used for bringing the metal to the bevel that you want. To smooth the metal and remove the grooves left by fast cutting, a technique known as "draw-filing" is employed. In draw-filing, the file is drawn obliquely along an edge or surface, rather than pushed across it. Smooth files are used in draw-filing, hence their name, since they smooth a surface.

Be careful when testing with your fingers an edge that you have been filing. A sharp file cuts so efficiently that you really can slice yourself on an edge you've been working on. When learning to file, always file away from an edge, never toward it; you don't want your hand or wrist to be cut if the file slips.

If a file doesn't seem to be cutting well and isn't obviously dull, try working it from slightly different directions until you get that good feeling of the file's teeth biting into the metal. Don't be bashful about bearing down—that's what makes the file cut—but make sure the work is solidly clamped. A beginner can have as much trouble getting the work clamped up tight and keeping it securely clamped as with the actual filing; this is why I recommend starting with the convenience of a real vise.

Clean the teeth of the file with a file card (a special kind of wire brush) from time to time while cutting and before putting it away. It helps, too, to tap it gently on the bench or on a wooden block every few strokes. Although rust is bad for files, don't ever put oil on them; it clogs their teeth. On the bench and in storage, keep files from banging against one another or against other tools. More files are dulled through careless handling than by cutting.

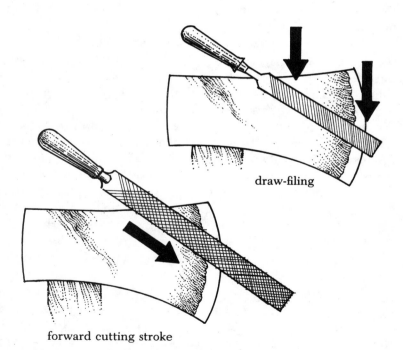

draw-filing

forward cutting stroke

**Figure 6-5:** The forward cutting stroke quickly brings an axe to the bevel you want; draw-filing smooths the surface, removing marks left by forward cutting.

# Sharpening an Axe

Now that you know enough about filing to get started, you're ready to sharpen a new axe. Axes are often sold with a general purpose bevel, a compromise that is not the best bevel for cutting logs of any thickness. If you want to learn to buck with the axe, the edge needs to be brought to a more gradual bevel, which means you must alter the shape of the edge as you get it from the store. Otherwise, you will be unnecessarily discouraged from bucking with the axe at the very beginning.

As with any sharpening job, it is essential to know before starting how you want the finished edge to look. Figure 6-6 shows an actual cross section to use as a guide in shaping the edge of an axe to be used for cutting through logs.

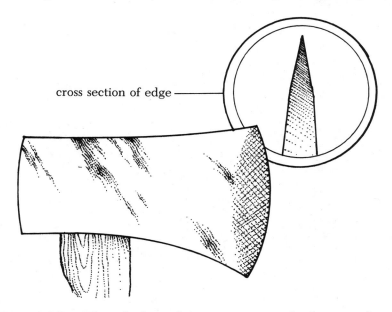

cross section of edge

**Figure 6-6:** When looking down on a properly sharpened axe blade, it shows a slight convexity. The bevel of a new axe usually has to be filed or ground to get a uniform cross section like this.

Two files will be needed for sharpening the axe, a 12-inch double-cut bastard mill file, and an 8-inch single-cut smooth mill file. The double-cut file will be used to take metal off to reshape the edge; the single-cut smooth file will be used to draw-file the edge smooth once it has been brought to the desired shape. Be sure to work with a good handle on each file.

To hold the axe for cutting with the double-cut file, clamp it firmly in the vise with the handle of the axe down (as shown in Figure 6-7). If your hands aren't particularly strong, put some body weight onto the vise handle when you tighten its jaws on the axe; the axe must not be able to move when you begin filing!

Now remove enough metal with the larger file to bring the edge to the correct shape. Take your time and make sure the file bites well on each stroke.

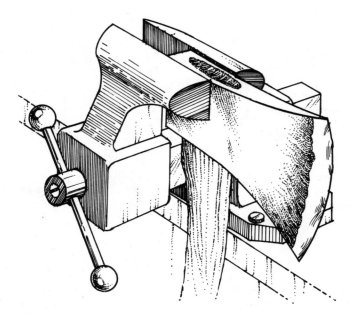

**Figure 6-7:** Axe in vise, ready for filing.

Once the edge has been shaped, it is ready for draw-filing to remove the marks left by the double-cut file. Clamp the axe now with the edge up, handle parallel to the floor, and draw-file the edge until you can't see tooth marks from the larger file on either side of the blade. Watch your fingers and hand while draw-filing!

The draw-filing will leave a smooth, attractive surface with only fine filing marks. Some believe it unnecessary to hone an edge that has been draw-filed, but you should learn how to give it the full treatment. Keeping the axe in the draw-filing position, take the sharpening stone, apply mineral oil generously to its coarse side, and hone both sides of the axe edge with a rotary motion. Then do the same thing with the fine side of the stone. Be careful not to cut your hand.

The oil is necessary to keep the pores of the stone from being clogged by fine particles of steel, so keep the stone well oiled while you work. This part of the job is a bit messy; an apron and a pair of rubber gloves will keep you clean. Keep a rag or two handy to wipe the axe blade and, when you're finished, the stone. A small brush is useful for applying the oil to the stone.

A round axe stone, which also has a coarse and a fine side, is even handier for honing an axe blade, particularly in the woods. I have one that I bought some years ago from Sears; unfortunately, I don't see it listed now by any of the suppliers I'm familiar with, but perhaps the handy round axe stone will come back now that axes are becoming popular again. In the meantime, the rectangular bench stone works well, and has the advantage of being useful for sharpening many other tools.

With a little patience and much care, you should be able to sharpen a pencil with your axe after your first attempt at hand sharpening. Be careful, in fact, not to overdo things; this approach is so effective that it's easy to produce an edge that is too fine for an axe, at least too fine for a beginner's chopping skill.

Nicks in the blade are removed by leveling the edge, then reshaping, draw-filing, and honing again with the sharpening stone. Deep nicks can require a lot of filing to remove them; the trick is to avoid them in the first place by being careful with the axe!

This isn't the whole story about sharpening axes, but it's enough to get a beginner well and safely started. For a more advanced treatment, read Alex Bealer's *Old Ways of Working Wood.* He provides the best discussion I know of on sharpening axes in the woods.

# Sharpening and Reconditioning Wedges

The same equipment used to sharpen your axe can be used to sharpen and recondition your wedges (see Figure 6-8). The large double-cut bastard mill file will take off mushroomed edges around the head of the wedge, and used with the smaller single-cut smooth mill file it will keep the cutting edge of the wedge sharp. Honing with the sharpening stone isn't necessary for wedges.

The only real problem in filing a wedge is watching out for your toes when clamping it up in the bench vise! A falling wedge can do a lot more than pare a toenail, so be careful, and don't let it slip out while you're tightening up the vise. Keep your feet out from under your workbench, just in case.

A bystander may tell you that it's quicker to grind a wedge down on an electric grinder than to reshape it with a file. True—but it's also more expensive, since you can wear a grinding wheel down fast reshaping a large wedge, and wheels are not cheap. If the wedge is properly clamped and you employ a sufficiently large, sharp file, it isn't that much slower to do an elegant job of filing by hand, and with no appreciable wear to the file.

When *will* a file wear out? I don't know, but I've got some pretty old files that are still a pleasure to use. I'm

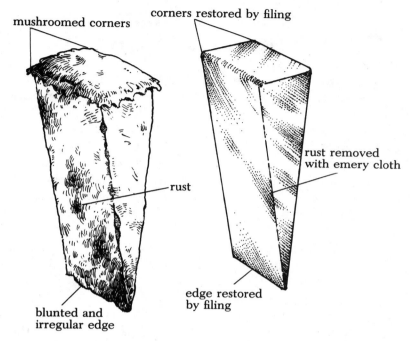

**Figure 6-8:** Reconditioning a damaged wedge.

intrigued, however, by a statement in the 1932 edition of the *American Machinists' Handbook:*

> One who has given the matter careful attention, and has built file-testing machines, Edward G. Herbert of Manchester, England, has come to the conclusion that a file does not cut best when it is new but after it has been used for some little time, say 2,500 strokes or the filing away of one cubic inch of metal. Another curious feature is that its usefulness seems to come to a sudden instead of a gradual end.
>
> —Colvin and Stanley,
> *American Machinists' Handbook*

# Touching Up and Resharpening Hand Saws

Now, you're ready for the big time—touching up and resharpening your woodcutting crosscut saw. The problem now is not to understand how to use a file, but to understand how the saw cuts. It's much easier to understand what you have to do if you begin by touching up the saw before it gets noticeably dull. You want to preserve as long as possible the shapes of the teeth and the way they are sharpened when the saw leaves the factory. If you don't wait until wear has created a restoration problem, you can easily see the bevels you're supposed to maintain.

So let's take a good look at the teeth of a woodcutting crosscut saw. Figure 6-9 shows the two kinds of teeth. The teeth that are beveled on their sides are called cutting teeth; the others are raker teeth.

The cutting teeth work by cutting across the fibers of the wood like pairs of knives. This is why they're beveled to make the knifelike cutting edges. And they're sprung

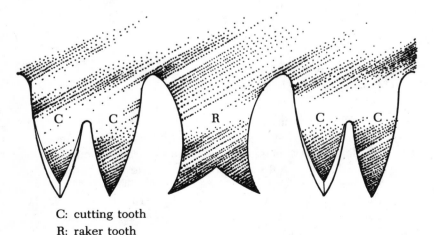

C: cutting tooth  
R: raker tooth

**Figure 6-9:**　Teeth on a crosscut saw.

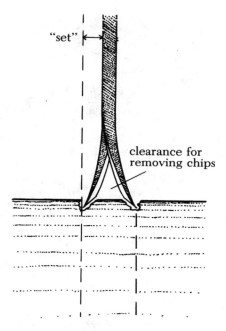

**Figure 6-10:** The actual "set" of the cutting teeth on a crosscut saw, shown here in cross section, is about half the thickness of the blade.

slightly apart, or "set" to keep the chips from jamming between them (see Figure 6-10). All saws designed to cut wood across the grain have their cutting teeth beveled and set more or less in the same way, even the fine-toothed little dovetail saw used in cabinetmaking.

The raker teeth are peculiar to the large-toothed wood-cutting crosscut saws. Bucksaws don't have them, and they are not found on carpenters' crosscut saws. The function of the raker teeth, as shown in Figure 6-11, is simply to clear chips from the groove made by the cutting teeth. Raker teeth have no side bevel or set, and are just a mite shorter than cutting teeth.

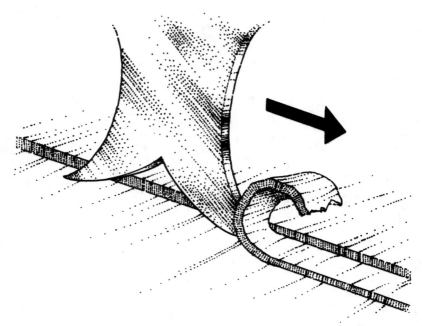

**Figure 6-11:**   A raker tooth on a crosscut saw serves to clear chips from the groove made by the cutting teeth.

We've been considering a saw that is sharp and cutting well. Figure 6-12 shows what happens to the teeth when a saw gets dull. The tips of the cutting teeth are worn down and their knifelike cutting edges are blunted, too. As the cutting teeth are shortened by wear, the raker teeth are pushed into doing what they are not intended to do, and their chisellike cutting tips become rounded.

Once you know what to look for, it's easy to see the tips of the dulled tooth points catching the light when you hold a saw upside down. But you don't even have to look. Your fingers can feel the difference, and so will your back and arms if you labor with a saw that chews the wood instead of slicing through it.

The best time to touch up a saw is after you have cut with it for only a few hours, before there is any noticeable

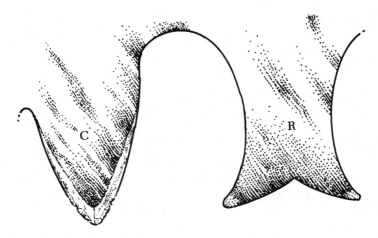

**Figure 6-12:** Dull teeth on a crosscut saw.

dullness. The job will then take only ten minutes. Here's how to do it:

1. Clamp the saw in a woodworking vise with the teeth up and with the saw handle to your left as you face the vise (see Figure 6-13).
2. With a crosscut file, touch up the bevels on the teeth bent toward you. One stroke of the file for each bevel should be sufficient. Start with the teeth near the handle of the saw and work toward the front of the saw. You may find it easier at first to do all the front bevels, then go back and do the rear bevels. Raker teeth do not need touching up until they are noticeably dull, or until the cutting teeth have been lowered to the same height as the raker teeth. You'll have to shift the saw several times to keep the area where you're filing free of vibration. Be careful not to cut your hand while shifting the saw in the vise. Keep a piece of chalk handy to mark where you stop and to mark teeth needing a bit of extra work. You may find it helpful at the beginning to check with a chalk mark the teeth you have finished touching up.

3. Turn the saw around in the vise, putting the handle on your right this time. Touch up the bevels on teeth bent toward you as you did before.

The larger two-man crosscut saw is touched up the same way, but mark one of the handles or one end of the blade so that you don't get confused when turning the saw around in the vise. If you find the larger saw unwieldy to handle without additional stiffening, cut two thin planks to the general shape of the blade and clamp them to the blade, then insert the whole assembly in the vise.

Do this touching up while the saw is still new, and you'll soon be so familiar with its teeth and how to file them that complete resharpening will be easy. You can do a first-rate

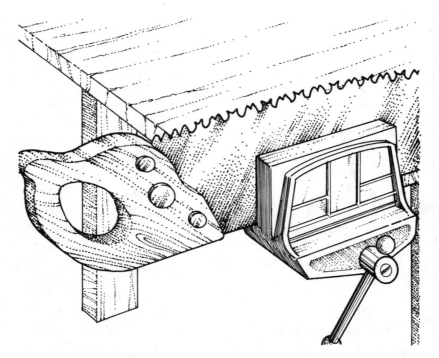

**Figure 6-13:** One-man crosscut saw in woodworking vise, ready for retouching.

resharpening job if you prepare in advance. Improvise a height gauge for the raker teeth from half of a brass door hinge and an automotive "feeler" gauge, as shown in Figure 6-14. Make a setting block from a piece of hardwood or metal beveled to match the set of the cutting teeth while the saw is new. To use such a block, lay it flat on the workbench, place the saw on the block, and strike every other cutting

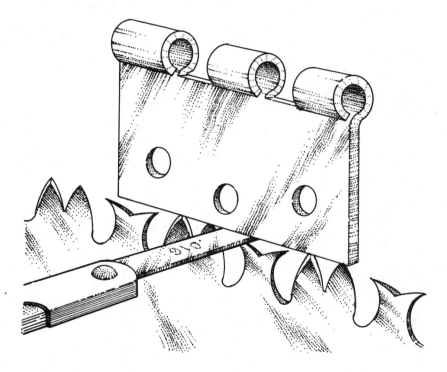

**Figure 6-14:** Improvise a height gauge for raker teeth by resting half of a 3½-inch brass door hinge against the tops of the cutting teeth, as shown. With an automotive feeler gauge, select a leaf that equals the difference between the height of the cutting teeth and the raker teeth. Record the information when the saw is new so you will know how much to touch up the raker teeth after you have used the saw for a while.

tooth with a small ball-peen hammer and a ¼-inch punch, working in the direction of the original set (see Figure 6-15). The raker teeth have no set.

If you acquire a used saw, you can still make the gauge and setting block to touch up and sharpen your saw. Get the dimensions from teeth close to the handles, or follow Reginald D. Forbes's recommendations from the section on sharpening crosscut saws in the *Forestry Handbook*.

For flat-ground as opposed to taper-ground blades, file raker teeth to 1/64 inch lower than cutting teeth; set or bend

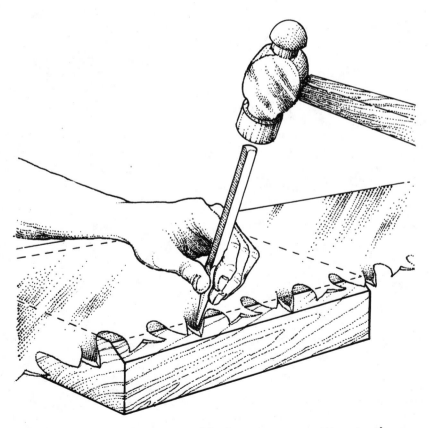

**Figure 6-15:**  Use a setting block to restore cutting teeth on a crosscut saw.

cutting teeth no farther than ¼ inch from the point of each tooth. Forbes says, "The amount of set for a modern taper-ground crosscut should be about 0.016 in. for softwoods and about 0.008 in. for hardwoods. For flat-ground saw blades, it should be about twice as much." Taper-ground means that the teeth of the saw are of thinner metal than the upper body of the saw. An old saw is more likely to be flat-ground. If you have trouble deciding whether your saw is taper-ground or flat-ground, ask any serious mechanic or metal-worker. The necessary measurement can be made in just a minute or two with a micrometer or vernier caliper.

Complete resharpening of a woodcutting crosscut saw, a bucksaw, or any saw used on wood, involves the following steps:

1. *Jointing the teeth* —bringing them all down to the same height

2. *Shaping the teeth* —refiling them to their original shapes

3. *Setting the teeth* —restoring the original set of the cutting teeth

4. *Sharpening the teeth* —restoring the original bevels

5. *Side-dressing the teeth* —removing slight burrs and irregularities left by sharpening

I'll just comment on these steps in outline, since you won't need to resharpen your saws for a long time if you touch them up regularly, and good detailed instructions are easy to obtain.

Jointing is done with a single-cut smooth mill file that is simply run along the points of the cutting teeth as shown in Figure 6-16.

On the bucksaw and other saws smaller than the woodcutting crosscut saw, setting can be done with a plierlike tool

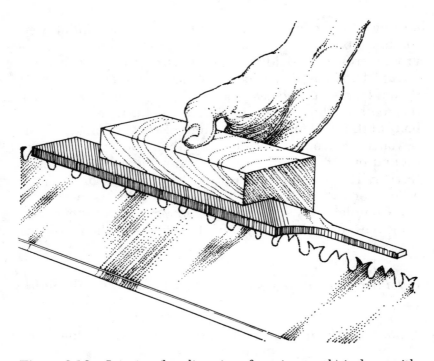

**Figure 6-16:** Jointing (leveling tips of cutting teeth) is done with a mill file.

called a hand saw set, which is adjustable for teeth of different sizes. Tapered saw files, triangular in cross section, are used to touch up and resharpen such saws. Side-dressing involves no more than passing an oil stone lightly along the sides of the teeth, from the back to the front of the saw, after the teeth have been sharpened.

To touch up a bucksaw blade, remove it from the saw frame and clamp it between two strips of wood, allowing the teeth to protrude about ¾ inch. Then:

1. Clamp the whole assembly in the bench vise with the teeth up and pointing to the left (see Figure 6-17).
2. Holding a tapered saw file at an angle of 70 degrees,

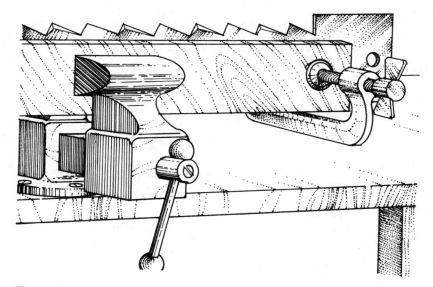

**Figure 6-17:** Bucksaw blade in bench vise, ready for retouching. To make filing easier, stiffen the blade with wood strips held by C-clamps.

with the point of the file directed toward the rear of the saw blade, work along the blade from left to right. File simultaneously the front cutting edge of each tooth bent toward you and the rear cutting edge of the adjacent tooth bent away from you. Figures 6-18 and 6-19 show two views of this step.

3. Turn the blade around in the bench vise and repeat the process.

Many older books on hand woodworking contain excellent instructions on the hand sharpening of saws. Modern books that present the details of sharpening woodcutting crosscut saws, bucksaws, and other saws are Alex W. Bealer's *Old Ways of Working Wood,* Reginald D. Forbes's *Forestry Handbook,* and Harry Walton's *Home and Workshop Guide to Sharpening.*

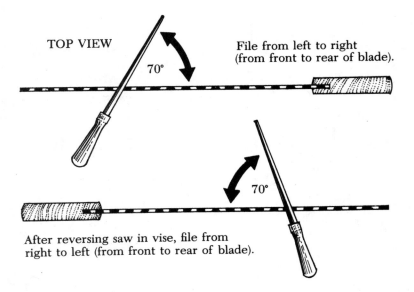

TOP VIEW

70°

File from left to right
(from front to rear of blade).

After reversing saw in vise, file from
right to left (from front to rear of blade).

70°

**Figure 6-18:** Retouching or sharpening the bucksaw.

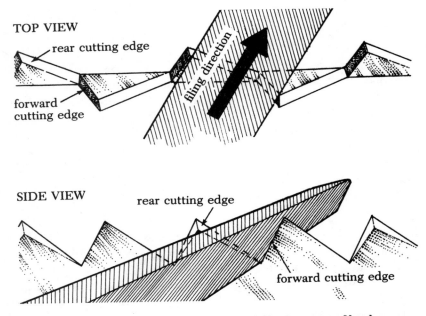

TOP VIEW

rear cutting edge

filing direction

forward
cutting edge

SIDE VIEW

rear cutting edge

forward cutting edge

**Figure 6-19:** In sharpening the bucksaw, the taper file sharpens
the *forward* cutting edge of one tooth and the *rear* cutting edge
of the adjacent tooth at the same time.

Bealer, as usual, is good on traditional methods of sharpening and the use of simple accessories. Forbes shows a variety of tooth patterns for woodcutting crosscut saws and the use of manufactured gauges and jigs for sharpening such saws. Walton's little book is a complete treatise on sharpening all kinds of tools.

# A Sheath for Your Axe

Though it is not uncommon for good axes to be sold without sheaths, you should have a sheath to protect yourself, the axe, and other people. If your axe comes without a sheath, make a sheath as soon as possible.

The best material is cowhide, which is available in local leather shops. Before buying the leather, make a pattern out of heavy paper or light cardboard. Even if you have never done this kind of work before, it shouldn't take more than a few attempts to construct a pattern, stapled together with ordinary paper staples, that fits your axe well. Be sure to allow sufficient material for folding and riveting (the leather will be about ⅛ inch thick, and a ⅝-inch margin is needed for riveting). If you want to be really sure of the fit before cutting the leather, try your pattern out first on a piece of indoor/outdoor carpeting—which, by the way, can be used as an inexpensive substitute for leather in the sheath itself.

When you've got the pattern sheath fitting right and the leather laid out, open up the pattern and transfer it onto the smooth side of the leather with carbon paper or a soft pencil. The *outside* of the pattern goes against the *smooth* side of the leather when tracing. Cut the leather with a sharp knife, backing the cutting with a pine board. Dampen the leather slightly from the smooth side, fold it around the axe, and mark for the rivet holes (see Figure 6-20).

On this kind of sheath, copper belt-rivets are a strong and very attractive way of holding the sections together. These rivets, the snap to close the cover of the sheath, and a simple tool for inserting the snap can be obtained where you buy the leather.

The holes for the rivets can be made with a nail, cut straight across with a hacksaw, and backed with a piece of pine, or you can make them with a leather punch. Here's how to do the riveting after making the holes:

1. Push the rivet through both edges from the rear of the sheath.
2. Place the copper washer over the shank of the rivet where it comes through at the front of the sheath, and push the washer down on the shank of the rivet as far as you can.
3. Check how far the shank extends beyond the washer with the washer pushed down as far as it will go. The shank should still protrude for a distance equal to half of its diameter; a little more is all right, but don't have it less. If the shank extends too far, cut it back to the correct distance with a pair of nail-cutting nippers.

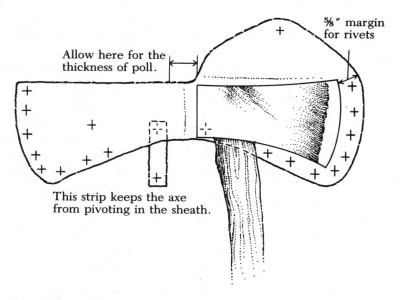

**Figure 6-20:** Pattern for an axe sheath made from leather or indoor/outdoor carpeting.

4. With the ball (rounded) end of a small ball-peen hammer, upset the end of the shank with light glancing blows all around, away from the rivet. Use the axe head as an anvil under the machine-made head of the rivet. When you have thus deliberately mushroomed the end of the shank so that the washer won't slip off, strike a few sharp blows with the squared face of the hammer straight down on the mushroomed second head you have just fashioned. This will draw the leather together firmly and tighten the rivet forever. Don't strike too hard, just enough to tighten everything up.

The ball-peen hammer called for here is not a special tool; you probably know it already as a "mechanic's" hammer, since it is used in automotive repair work and light metalworking. One with a 4-ounce head would be just right for this job, though the next size, with an 8-ounce head, could be used, too.

Copper is pleasant to work with. I haven't specified the rivet size, since the leather comes in slightly different thicknesses. Fold the leather double in the store, and you'll be able to tell whether the rivets are sufficiently long to secure the sheath.

Riveting as I have described it is really quite easy. This method goes back to horse-and-buggy days, and isn't always described in contemporary leatherworking manuals. There is a simple tool called a rivet set which is driven by a hammer to force the washer down the shank of the rivet and head the rivet; this tool, too, is sold in leathercraft shops. Direct hammering produces a slightly irregular second head, but the effect is more interesting, hence my suggestion that the machine-made head of the rivet be inserted from the rear of the sheath.

If riveting seems a bit much to get into, there are other ways of making sheaths that involve simply lacing or sewing the leather together. Excellent detailed instructions for making sheaths this way can be found in Ben Hunt's *Complete How-to Book of Indiancraft.*

Hunt was a fine craftsman and woodsman, and an expert in adapting North American Indian craft techniques. The book just cited also has good plans for a homemade pack frame and for homemade snowshoes. Another of Hunt's books, *The Golden Book of Crafts and Hobbies* contains more ideas for homemade woodcutting equipment: a heavy-duty sled and a trail toboggan. In all of Hunt's work, only the simplest tools are called for, the designs are practical, and the instructions clear and beautifully illustrated.

# Replacing an Axe Handle

If you are careful with your axes and sledgehammers, it will be a long time before you break a handle. Nevertheless, it is smart to have an extra handle ready for your favorite axe before it is needed.

Replacing an axe handle—the woodsman's term is "hanging an axe"—is not difficult, but like saw sharpening, takes a bit of patience. Getting the stump of the old handle out of the eye of the axe should be done carefully, too, particularly if metal wedges, as well as a wooden wedge, have been used to hold the handle in the eye of the axe. I've never tried the old stunt of burying the axe blade in the ground with the eye and poll (the flat end of the axe) protruding, and building a fire over the axe head. My own more prosaic method is to cut the stump off close to the head and clamp the axe head in a bench vise. Then with a ¼-inch metal-cutting bit chucked into a hand drill, I drill a number of holes through the wood of the stump around the metal wedge. With the wood thus honeycombed, it is easy to tap it out with a small cold chisel and a ball-peen hammer. The reason for using the metal-cutting drill bit and cold chisel is to avoid damaging woodworking tools when removing the wood around the metal wedge.

The next step is to work the top of the new handle down with a rasp until it fits the eye of your axe. An inexpensive caliper is a great help at this stage. Notice that the opening at the top of the eye is larger than at the bottom; this is to allow for the expansion produced by the wedge driven into the slit in the top of the handle.

When you have begun to get a good fit, cut off the tip of the fawn-foot at the end of the new handle (see Figure 6-21) so that you can drive the handle home without splitting it.

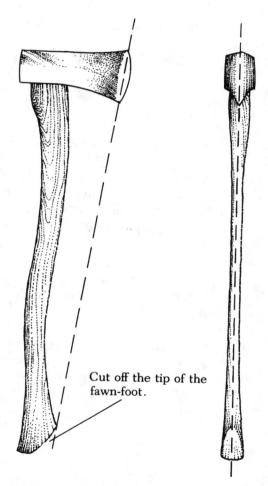

Cut off the tip of the fawn-foot.

**Figure 6-21:** Proper alignment ("hang") of an axe.

While trimming the new handle to fit the eye of the axe, keep track of the alignment of the handle with the head, since that affects how accurately you can cut with the axe. These instructions for checking alignment are useful:

> Sight along the back of the helve [handle] to see if it is straight in line with the eye of the axe, then turn it over and see if the edge of the axe ranges exactly in line with the center of the hilt [rear end of the handle], as it should, and that the hilt is at right angles to the center of the eye.
>
> —Kephart, *Camping and Woodcraft*

Forbes's *Forestry Handbook* has illustrated step-by-step instructions for hanging an axe.

The lore of axe handles is part of the great tradition of the axe as a tool; in the past, woodsmen and woodcutters preferred to make their own handles. Bernard Mason tells how to make an axe handle by hand in his *Woodcraft and Camping.* David Grayson devoted a whole essay to the making of an axe handle in *Adventures in Contentment.* It's a good piece and worth reading, though if you've read Christopher Morley's delightful parody of Grayson in *Parnassus on Wheels,* you may find it hard to concentrate seriously on anything in *Adventures in Contentment.*

These are real old-timers for most people today. Robert Frost (I hope) is not.

> He showed me that the lines of a good helve
> Were native to the grain before the knife
> Expressed them, and its curves were no false curves
> Put on it from without. And there its strength lay
> For the hard work.
>
> —Frost, "The Ax-Helve"

His poem is, of course, about much more than the making of fine axe handles. But there is technical as well as human wisdom in it, and it can help you when selecting an axe handle.

# Chapter 7

# Transporting and Storing Firewood

It was a cord of maple, cut and split
and piled—and measured, four by eight.
And not another like it could I see . . .

—Frost, "The Wood-Pile"

There are several important reasons for giving advance thought to how you are going to transport and store the wood you cut. By planning your traveling to and from cutting sites, your loading and unloading procedures, and your storage arrangements, you can prevent injuries from moving and handling wood.

Another important reason for planning is to avoid waste. Firewood, even "free" wood, is valuable, and not only in a strictly monetary sense. There are too many parts of the world where people must walk all day to gather a fraction of the wood many of us can still collect not far from our homes. It just isn't right to let wood go to waste because we cut more than we can sensibly transport, or because it has been stored carelessly after being brought home.

Finally, when you're fortunate enough to be able to locate, cut, and bring back a good load of wood, it's helpful to know before you come up the driveway late on a cold afternoon how you're going to unload your precious cargo. It's not the worst thing in the world to have to rearrange your wood storage area during the season, but there is, after all, plenty of other agreeable outdoor woodcutting work to do around the house.

# Street Logging

The most romantic way to get your firewood is from a spot off the road, in a forested area. This is also the most difficult way, requiring experience if the wood is to be gotten out safely and efficiently. Let's start, then, with a seemingly more prosaic situation—salvaging wood close to home that would otherwise go to the dump. It not only makes good sense from the standpoint of economy to collect such wood whenever you can do so legally, it's also an excellent way of preparing yourself and your equipment for off-the-road logging. However, before I get into actual methods of handling and transporting logs and limbs, you will want to be aware of other factors to keep in mind when "street logging."

First, there is the question of legality. Be sure you're aware of local ordinances and practices concerning people, other than city or village crews, picking up material of any kind set out for collection.

After severe storms—when a great quantity of wood may be available for the taking—there are a number of additional considerations. Watch out for downed power lines. Be careful not to interfere with work by emergency crews. Unless you have explicit permission from property owners, don't disturb limbs that have damaged cars, fences, or other structures; the owners may be waiting for insurance adjusters to survey the damage.

On city and commercial property, ask to take the wood.

Seek permission from police, supervisors, or other people close to the scene. If you appear to know what you're doing and are open enough to ask, you'll probably get a friendly go-ahead; at the very least, you'll spare yourself any embarrassment.

Finally, secure your load as carefully for a short run as for a longer trip, and flag long loads on your car, truck, van, or trailer. Careless preparation is unsafe, and it may be illegal.

As I suggested previously, even when you're working right down the street from your house, there are considera-

**Figure 7-1:**　A chain can be secured around a log with a grab hook.

tions comparable to those that apply in woods miles away. Trunks and limbs have to be *cut,* heavy log sections have to be *moved* to your vehicle, and the wood you're going to transport has to be *loaded* and *secured.*

The cutting comes under the headings of *limbing* and *bucking,* which I will discuss in the next few chapters. I'll talk about loading, securing, and hauling now, since a log on a city street is just as heavy as one in the woods.

The most important thing to emphasize is that wood is heavy. Large logs and heavy limbs should not be carried to your vehicle, nor should they be loaded by being heaved clear off the ground and tossed into your car, van, or truck. *Drag* them with the aid of a chain or piece of rope. Chain

**Figure 7-2:** A timber hitch knot is good for hauling a single log off the site.

(one ⁵⁄₁₆ × ³⁄₁₆ inch is quite sufficient) is fastened around the log with the help of a grab hook (Figure 7-1).

If you are using rope, get a length that is ⅜ inch or heavier. Bend it once around the smaller end of a log and secure it with a timber hitch (Figure 7-2).

To load, first raise or lift one end of the log until it rests on the edge of the cargo area of your vehicle. Then raise the other end carefully, pushing forward if you're loading a truck or trailer, or pushing forward and then swinging the log around if you're loading into a van or car trunk. Remember to bend at the knees—not from the waist—when lifting the log.

This assumes you're working with stuff that isn't really heavy. Very heavy logs should be handled by two people, or, if you're working alone, with the aid of a block and tackle.

# The Block and Tackle

The block and tackle is a labor-saving device which reached its maximum development in the days of sailing ships. I have two sets, a large one which takes ½-inch-diameter rope, and a smaller one which uses ⅜-inch-diameter rope.

A ⅜-inch-diameter Manila rope has a safe working load of 120 to 390 pounds, depending on its age and condition; ½-inch-diameter Manila rope has a safe working load of 250 to 800 pounds. Manila rope is cheaper than nylon, and wears very well when used properly. (Several special knots and slings have been developed for hoisting heavy cylindrical objects like logs, but those shown in Chapter 6, "Caring for Tools and Equipment," will work well on the site.)

The small block and tackle has two pulleys above and one below. The traditional term for such an arrangement is "watch or single-luff tackle"; it has a theoretical mechanical advantage of 3 to 1. This means that, discounting friction, your pulling effort on the free end of the rope is multiplied three times. When you hang the block and tackle from the roof of a vehicle or a support of some sort, and pull down, even a light person can easily lift an object weighing more than 100 pounds.

My larger block and tackle has three pulleys in the upper block and two in the lower block. The old nautical term for this combination is "double-luff tackle," and its theoretical advantage is 5 to 1. If you exert a force of 100 pounds downward on the free end, you can—theoretically —lift an object weighing 500 pounds, or ¼ ton. Calculations made for the losses due to the friction of the ropes in the grooves of the pulley bring the actual mechanical advantage down to 3.3 to 1. In the case of a single-luff tackle, the actual mechanical advantage is about 2.3 to 1.

A one- and two-pulley block and tackle with ⅜-inch rope, like my smaller block and tackle, is all you need for a van (see Figure 7-3). Make sure you have as strong an upper

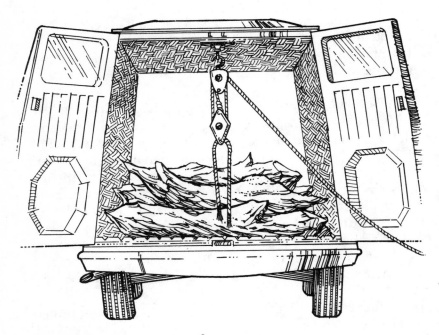

**Figure 7-3:**   A hoisting rig for a van.

door catch as I have in my Chevrolet van before you begin hoisting logs with a block and tackle. I hung on mine with my feet off the ground to test its strength.

When I owned a pickup truck, I made a rig (pictured in Figure 7-4) from salvage materials and bolted it to the rear of the truck. With this larger block and tackle, I could handle logs 8 feet long and 12 inches in diameter. This kind of hoisting can't be done with a car, but then, for the sake of the vehicle, a car shouldn't be loaded with extremely heavy logs anyway.

To prepare a block and tackle for use, thread the pulleys with rope. I used 25 feet of rope on my smaller combination. There is a place to tie the pulley on the lower block. The upper block—the one with more pulleys—is hooked to your support. To the lower block, the movable one, an extra

**Figure 7-4:** A hoisting rig for a pickup truck.

length of ⅜-inch rope, about 7 feet long, is attached; there
will be a hook, ring, or slots for this purpose. This rope
should be doubled so that you have two equally long ends
to use for securing the object to be hoisted.

To hoist a log section, fasten each of the free ends of the
7-foot rope to the log with two half hitches, spacing the
hitches so that each is about 6 or 8 inches from the middle
of the log section. With practice, you can hoist with one
hand on the *fall*—the end of the long rope to which power
is applied—while you guide the weight being hoisted away
from the bumper of your vehicle. At the beginning, have a
companion guide the log as it goes up so that you can concentrate on pulling and keeping everything moving
smoothly.

This is an old-fashioned, but very effective, method of
lifting heavy weights. Take it easy at first, not because it
requires much strength, but because it will take time to
learn to manage the ropes. Keep your feet out from under
the weight at all times and watch your hands when guiding
a log up and down. It's so easy to hoist a heavy log right off
the ground that you are liable to forget how much weight
you are working with. I don't want you to be reminded by
a log mashing your hand against a bumper, or coming down
on your fingers.

Learning to use a block and tackle is a great way to
spare yourself back strain. Practice first with light pieces,
then move on to the bigger stuff. If you are twenty-five years
old and in top condition, you may not want to bother with
such a labor-saving rig, but learning how to use it will enable
you to safely handle heavy logs well past middle age.

# Securing and Hauling a Load of Wood

How much of a load can *your* vehicle handle? This
depends not only on its size, but on its construction. I would

be especially careful with newer, light vehicles. You *know* you're overloading when the rear end sags close to the ground. Unless you've successfully done this kind of thing before with that vehicle, remove part of the weight. An overloaded vehicle is subject to a lot of expensive damage; more important, it's hard to drive safely.

If you don't have a pickup truck, van, or car with a lot of cargo space, a sturdy two-wheel trailer is a great help, since any vehicle can pull much more than it can carry. The trailer can be very simple (I have seen many ingenious home-built cargo trailers, made from salvaged auto parts), but it should be legal, and with springs, wheels, and tires that can handle the load.

The trailer should be attached to the rear of your vehicle with safety chains, as well as by the trailer hitch. For short local hauls at low speeds, ½-inch rope can be used to secure logs to the trailer frame or bed. Longer trips at highway speeds require the use of chains crossing the load of logs at two or more points. To tighten the chains, use a chain tightener on each chain; this is a lever-action device that takes up the slack in the chain. Chain tighteners are available from hardware dealers serving tradespeople and contractors' suppliers. Your logging chain can be used for one of the tie chains; another length of ¾ × 1¼-inch chain should be used for the second tie chain. The smallest size chain tightener will be adequate for securing these chains. After levering the handle down against the chain to tighten it, tie the handle of the chain tightener to the chain with a short length of cord.

A load of logs projecting beyond the rear of your trailer should be flagged with a bright red cloth. If you are traveling after dark, there should be provisions for illuminating the rear of the load.

When you're fortunate enough to get a fine load of wood, drive at considerably reduced speeds. It's easier on your equipment, and much safer, since a heavily loaded vehicle, or one towing a loaded trailer, has markedly different driving characteristics. It steers and corners differently, and of course needs much more stopping distance.

With some successful street logging to your credit, you're ready to go farther afield. But if you are going into rougher country in winter, keep in mind that cold weather can put an intolerable strain on a feeble ignition system, and navigating country roads with a load can bring out other weaknesses; so make sure you're driving off with something you can trust.

# Cutting Farther Out and Off the Road

Off-the-road cutting sites certainly present more challenges than those sites close to where you can park your vehicle. Unfortunately, these trickier sites are the ones you'll probably be forced to turn to as easily accessible wood gets scarcer. The more experience you have working with the block and tackle rig I described earlier, however, the farther you'll be able to go from your vehicle and bring the wood back for loading.

Generally speaking, unless you have a good four-wheel-drive vehicle *and* considerable driving experience, it's better to park your vehicle where you know it can't get stuck, and drag the largest log sections you can manage back to it. It's time-consuming or expensive—or both—to be bogged down somewhere in the woods, miles from home, so be cautious rather than adventurous in taking any vehicle off the road, particularly where there is snow or mud.

Just how far you can sensibly go to cut wood depends on variables that you will have to judge personally. You may be dragging wood uphill or downhill. There may be boulders, streams, marshy spots, or other obstacles in the way between the cutting site and where you are parked. The kind of hauling equipment you have, whether you're working alone or with capable companions, how much time you have—all these factors must enter into your decision. Every situation is different. But that's what makes woodcutting a sport and an art, not a cut-and-dried affair.

The best procedure is always to play it safe and stay well within the bounds of what you are confident you can do. Base your confidence on experience derived in simpler situations, like street logging close to home and work you've done at roadside cutting sites.

There is another important point to keep in mind when using a vehicle in woodcutting. Living trees, and not just very young ones, can easily be damaged by vehicles. Running over roots, compacting earth around the roots, scraping bark—all these things, which can result from careless vehicle handling, are serious hazards to living trees. So, even if your vehicle can make it off the road into a wooded area, you'll want to assess these risks. It's better to haul your wood a short distance if driving off the road might do such harm.

Even if you're working at a roadside site, which usually includes cutting in a town or municipal dump, you'll probably have to do some dragging with a rope or chain. If you are working around living trees, be careful not to drag logs in such a way that you damage the trees. Bumping young roots or trunks with heavy logs, or scarring them with a chain, can be quite harmful to them.

Generally speaking, when working in the woods it's better to use your time to collect and load the biggest stuff you can handle, and then cut your logs down to burning length back home. But there are situations in which such cutting will be done off the road, so let's consider ways of getting the shorter lengths to your vehicle. The main thing is to avoid carrying wood for any distance cradled in your arms, particularly on uneven, rocky, or slippery ground. Carrying heavy loads this way is a severe and unnecessary strain on your back; there is also the danger of stumbling and taking a bad fall.

A homely but very effective solution, if the ground isn't too rough or snowy, is a wheelbarrow. New wheelbarrows, like just about everything else today, are expensive, but discarded ones are worth fixing up if they were once sturdy.

New handles can be made, a bit of welding or riveting can mend even large holes, and good wheels can be obtained secondhand from salvage dealers.

On snowy ground, a good load of cut wood can be hauled on a homemade sled or toboggan. Ben Hunt's book,

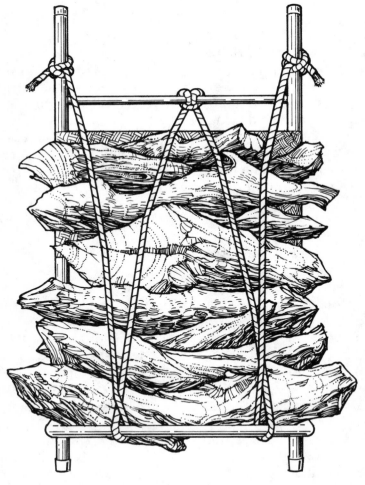

**Figure 7-5:** A recommended lashing pattern for a pack frame.

*The Golden Book of Crafts and Hobbies,* has good plans for making a heavy-duty sled and a trail toboggan.

If you have to *carry* your wood for any distance, get the load onto your back; it's less of a strain on your body that way, and you can see where you're putting your feet. This is a physically demanding way to get wood back to your vehicle, but it has its compensations: it'll keep you in shape for camping, and almost any load you carry in the summer will seem a lot lighter!

The backpacking arrangement I have used most is a guide pack frame (available from L.L. Bean) with extender bars. The load has to be lashed securely, even for a short walk, but a heavy load can be tied and untied very quickly with the simple arrangement pictured in Figure 7-5.

To stand up with a heavily loaded pack frame, follow these steps:

1. Sit down against the propped-up frame and get your arms through the shoulder straps.
2. Go forward from this sitting position onto your hands and knees.
3. Move one foot forward, raising your back carefully, then move the other foot, and stand up, using your thigh muscles. Don't be bashful about using a pole, sapling, or tree trunk to pull yourself erect. Such support is a great help.

Another way of carrying a heavy burden on your back is with a burden strap, which can be improvised from old web strapping. To carry wood this way, the log sections should be a little wider than you would want for lashing to a pack frame; if they're less than 2 feet, it's hard to tie and balance the load. If you don't want to carry log sections this way, keep it in mind as a way of packing out bundles of limbs and twigs for kindling.

People in developing countries still carry substantial burdens this way, passing the strap across the forehead in an arrangement called a tumpline (see Figure 8-6). Most of us would get a sore neck if we tried the tumpline strap method, but with just a few minutes of experimenting you can adjust straps or heavy rope by tying the ends across your chest (see Figure 8-7). This is also a good way of moving moderately heavy sections of wood when you're *unloading* your vehicle, if you don't want to drag them across your yard.

# Unloading and Storing Firewood for Seasoning

As much care should be exercised in *un*loading logs and cut wood as in loading; more so, perhaps, if it is dark, you're tired after an afternoon of cutting in the woods, and everyone is in a hurry for supper. It will be helpful if you have planned your wood storage in advance.

Wood should be stacked so that seasoning can proceed as well as possible. A sunny spot, with good air circulation, is preferable. In a temperate climate, finely split wood needs a minimum of six months to season adequately for burning; generally, eight months to a year is advisable. It is not difficult to tell when wood is seasoned sufficiently for burning. It will feel lighter and look aged, and there will be cracks visible at the ends.

Should you cover your woodpile? There are knowledgeable people who advise covering a woodpile with plastic or metal. I leave my woodpile naked to the elements; here are my reasons. First, I think more moisture gets trapped by covering than avoided. Also, once wood has been seasoned, rain or snow is not going to undo the process, but only wet the wood superficially. Finally, an open woodpile is a beautiful thing, one of the grandest forms of expendable art. Covering it with plastic (of all things) is sacrilegious—and messy, too, once wind, rain, and snow get to fooling with the covering. I would rather have my top pieces a little damp on the outside, since there is a simple cure for such a minor ailment. Get your fire going with the small supply of dry wood you keep inside, and let the slightly dampened wood from the woodpile toast a bit near the stove or fireplace before sending it in to do its duty.

As for a woodshed—an open, covered structure large enough to hold a season's supply of firewood—personally I don't see the need, the space, or the cash for one. A shed is a building that may be ruled on by town officials from the standpoint of conformity to building codes, and it may well affect your property taxes. Confer with an experienced contractor or an architect before undertaking this building project. Their fees are well worth the aggravation spared, the sturdiness and attractiveness insured.

# Preventing Wood Rot

Though I don't mind at all letting my woodpile go "bare headed" in all weather, I do insist that it be "dry shod." In other words, I'm more concerned about what's underneath a woodpile than what's on top. The most important thing is air. Unless the ground temperature remains below freezing, logs and wood placed directly on the ground will rot. There should be some arrangement, then, to keep the bottom of your woodpile from resting on the ground. Use bricks, cinder block, pipe, stones, discarded loading pallets—anything that permits air circulation under the woodpile.

# Damage to Dwellings from Wood Pests

You and I have a lot of bacteria in our innards, but not the kind that permit us to digest cellulose, the major solid ingredient of wood. Cows can digest cellulose, but they are not going to munch on your woodpile. Termites, carpenter ants, and powder-post beetles are another matter. The danger is not that they will eat you out of your firewood, but that they will literally eat you out of house and home. To be extra safe, observe these precautions while you're acquiring experience and finding out just what the risks are in your area.

1. *Keep your woodpile well away from your house and other permanent structures.* Never use a house wall, even a brick, stone, or concrete wall, as backing for a woodpile.

2. *Don't store wood on porches, or inside your house.* Yes, this means lugging wood in every day, but until you learn to tell the difference between sound wood and wood that may be infested, it is the most prudent course to follow. Carpenter ants and powder-post beetles can migrate from firewood conveniently stored by the living room fireplace into the wood of the house itself.

Just how great the likelihood is of infestation from termites and other cellulose-eating insects depends on a number of variables: whether you keep your wood from contact with the ground; how long you store wood before using it; the way your house is constructed; and various local conditions. Talk to local agriculture extension personnel, college and university entomologists, experienced contractors, and local architects for advice and up-to-date local information.

Do some reading. There is a good United States Department of Agriculture report on termites, *Subterranean Termites, Their Prevention and Control in Buildings,* by H. R. Johnston, Virgil K. Smith, and Raymond H. Beal. To answer your questions about carpenter ants and powder-post beetles, browse through Robert E. Pfadt's *Fundamentals of Applied Entomology.*

As always, there is *The Wood Handbook*—that book whose praises I never tire of singing. Here there is a thorough discussion of bark beetles, ambrosia beetles, round-headed and flat-headed borers, as well as of the more common wood pests, including very clear descriptions of the visible evidence of wood-attacking insects.

*The Wood Handbook* also discusses molds, stains, and decay. Molds and stains are of no significance for wood that is to be used as fuel. Decay, which does reduce the amount of combustible matter, is minimized by seasoning and storing wood properly.

But don't let any of these considerations frighten you. I'm not the only person who has not had the least bit of trouble after seasoning wood in backyards and lugging it inside for years and years. I've never seen a termite in my woodpile, never had a chair collapse under me from the clandestine mastications of powder-post beetles, and have never been bitten by a carpenter ant.

There are, I must admit, reptiles around my woodpile here in South Carolina. Not snakes, but slender and agile little lizards, make their home here. They're always welcome, because they are great cockroach hunters—and we spend a lot more time thinking about roaches than about any wood pests.

# Accessibility

One of the great pleasures of woodcutting is to have all that wood right there in the backyard so that you can enjoy working on it in the open air all week long. You'll need room for a sawbuck and chopping block. If you do a lot of work with an axe, allow, too, for a pile of chips after each cutting session. There will also be a good deal of bark to pick up.

Chips and bark are not trash! Large chips and large pieces of bark are good for burning when thoroughly dried; this means giving some thought to drying arrangements. Small chips and small pieces of bark should be saved for mulch. They can also be used for compost, though they take longer to break down than lighter organic material. And, of course, twigs and smaller limbs should be collected for kindling and quick fires.

Inside storage of firewood is a matter of personal taste, although it is not the best idea to have large amounts of wood sitting around the house for weeks or months at a time —at least not until you have become thoroughly familiar with the differences in appearance between sound and infested wood. Whatever you decide to do about inside storage, give some careful thought to how you are going to be getting your wood into the house, particularly at night and in bad weather. It would be regrettable to develop sensible ways of transporting wood from the field and storing it outside, then to slip and injure yourself on the back steps carrying an armload of wood into the house. To carry wood, use an old bucket, or make a rectangle of canvas with rope handles. A sling may be improvised from a few feet of rope or webbing. The important thing is to have a free hand, and a chance to watch where you're putting your feet.

Finally, even if you choose not to store wood indoors for long periods of time, that doesn't mean that there isn't an advantage in having close at hand a variety of fuel and kindling sizes at any given time. There's not much point in having a fine stack of seasoned logs cut to burning length if you don't have the intermediate stuff needed to get the big fellows burning. The intermediate stuff, cut from boughs and branches, has to be ignited in turn by still finer material such as fuzz sticks. (A fuzz stick is a slender piece of pine or other softwood, whittled into a number of long shavings that are left attached to the stick. Three fuzz sticks, each about 6 inches long, stacked teepee style with the shavings curling downward, make an excellent fire starter.) Twigs, dry chips, generous pieces of dry bark, and pine cones are also excellent kindling; the fragrance of the pine cones makes them nice to have around before you burn them, and just looking at them makes most people feel good.

# Chapter 8

# Limbing

When I began to reflect on what I was doing as an amateur woodcutter, I found it helpful to think in terms of five successive cutting operations. With experience, all can be done with just an axe, but when one is starting out, a few other hand woodcutting tools are very helpful, so I'll discuss their use, too.

The basic cutting operations for an amateur are:

1. *Limbing* —removing branches from the main stem (trunk) of a tree

2. *Bucking* —cutting up the trunk into logs of manageable length, usually no more than 4 feet long

3. *Sectioning* —cutting logs into lengths convenient for burning (12 to 18 inches long)

4. *Quartering* —subdividing the thicker sections longitudinally into four or more pieces

5. *Splitting* —dividing the "quarters" into thinner pieces, two or three to a quarter

Prominently absent from this list is *felling,* or bringing down a standing tree. I omit it because for beginners it is neither necessary nor safe to fell trees. To be sure, every chain saw owner's manual tells you how to fell a tree in a few paragraphs or a page or two, and there are other things in print—not, however, what is read by professional woodsworkers—that make felling sound like something a beginner can do if he or she will only check out this, that, and the other item.

In themselves, these instructions are usually okay. The trouble is, until you have worked around trees for a good while, you can't use such abbreviated information safely. There is a chapter on felling further on, but I have included it for information, not for use in the early stages of your woodcutting. When you have read it, you'll understand the reasons for my caution.

# Limbing— Your Introduction to Axework

Limbing is not difficult or complicated; nevertheless, it has a number of important purposes. First and most obviously, it clears the trunk so that it can be bucked into logs. Removing projections facilitates dragging the logs to a loading point; also, logs that have been neatly limbed are easier to load and store. Finally—and particularly important for the amateur—limbing makes limbs and branches available for fuel, too.

Limbing is not only first as a cutting operation; it is also an excellent beginning to the whole art of woodcutting, since it affords not too arduous an introduction to real axework, and provides intimate experience with the structure of trees.

The more you know about the nature of what you're working with, the more you can substitute skill for force— or wasteful "artificial energy." So, let's take a closer look (in Figure 8-1) at how the limbs you'll be severing grow out of the tree when it is alive.

You can see right away why, if the limb slopes toward the crown (most do in hardwood trees), you have more working room if you cut from the root, or butt, end of the trunk. But since you're also interested in learning to use an axe competently, take another look at the diagram and study the way the grain runs. Notice how the pith or core

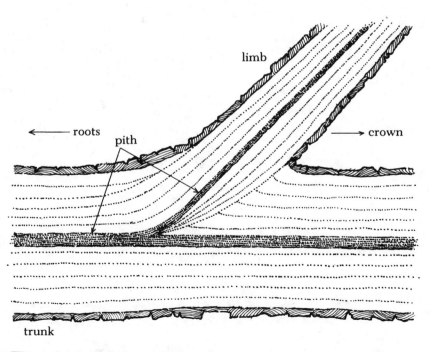

**Figure 8-1:** This cross section of a tree lying on the ground shows the way the pith and grain of a limb separate from that of the trunk.

of the limb departs from the pith of the trunk, and how the rest of the grain of the limb is related to the grain of the trunk.

Now what is an axe? A "chopping" tool? Not at all—if by "chopping" we mean what one does in a kitchen with a cleaver. An axe is a *slicing* and *splitting* tool. It works as it should when it cuts diagonally across the grain at the least possible angle and at the same time penetrates between the fibers of the grain as deeply as possible (see Figure 8-2).

What you should do, then, when limbing, is in each case to make a first cut, then a second cut parallel with the trunk to free the chip (Figure 8-3), repeating the sequence as necessary until the limb has been cleanly severed.

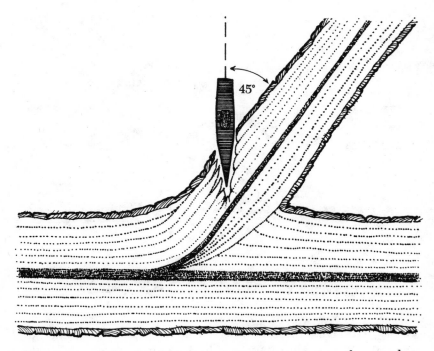

**Figure 8-2:** When limbing correctly, the axe enters the wood at a 45° angle from the bark, and penetrates the grain deeply.

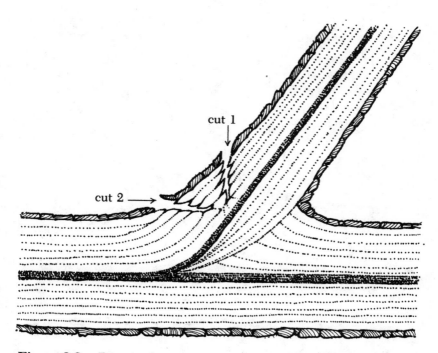

**Figure 8-3:** Repeat these two cuts in sequence until the limb is free from the trunk.

Since the angle at which the axe enters is so important, it should be evident why an axe has to be properly sharpened to cut well. An axe with too blunt a bevel will not be able to enter the grain at a sufficiently low, easy-cutting angle, and it will not be able to penetrate deeply enough for efficient splitting. Figure 8-4 illustrates the problem. If you try to make an axe that is too blunt enter at a better cutting angle, it will bounce, jarring unpleasantly, perhaps even slipping dangerously. Keep working on your axe until it cuts cleanly and effectively!

An axe with a 3½-pound head is more than adequate even for large limbs. But since care and accuracy are what are most important, and often you will be cutting at points

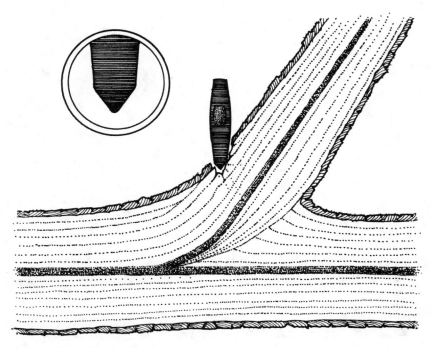

**Figure 8-4:** An axe with an insufficient bevel encounters too much resistance for effective slicing action.

several inches off the ground, you may prefer to work with a lighter axe (with about a 2-pound head) and a bucksaw, using the axe for smaller limbs and sawing off the big fellows.

The kinds of situations in which most limbing is done fall into two broad categories:
- the tree is resting on the ground
- the tree is in a pile pushed together by a bulldozer

The ground situation is more common and easier for a beginner to tackle. If a tree is resting on the ground, the trunk of the tree may be lying on the ground, *or* the trunk itself may be off the ground, supported by the root mass at one end and a few limbs at the crown end (see Figure 8-5).

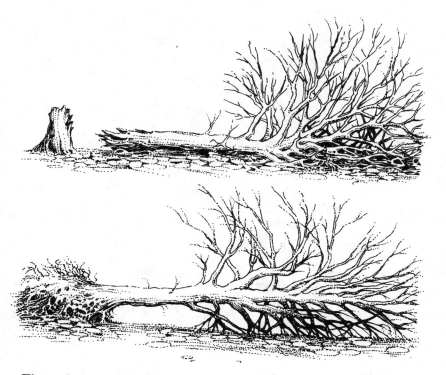

**Figure 8-5:**   Both trees here are resting parallel to the ground, but the one supported by root mass and limbs presents a more difficult situation for the woodcutter who wants to limb the tree.

The case where the trunk is lying along the ground is usually the easiest of all. You don't have to decide which limbs should not be removed at this time, you will not be cutting any limb more than a few inches from the ground, and there will probably be no shaking or vibration to contend with.

The procedure here is simply to work along the trunk from the root end toward the crown, severing the limbs as close to the trunk as you can. For safety's sake, cut across the trunk as much as possible, so that the trunk is between you and wherever you are cutting.

There are other important safety considerations:

1. *Always take the time to make sure you can swing your axe without any interference, even from the smallest twig or plant.* Such distraction can make you cut inaccurately and therefore hazardously.

2. *Check your footing continuously.* Whenever you change position, move your feet around a bit to make sure you are not standing on loose brush, slippery rocks, or anything that could shift underneath you suddenly.

3. *Make sure your clothing is right for this kind of work.* Too tight, and it will hamper your swing; too loose, or with anything dangling in the way, and it can make for an even more serious hazard. Here, too, take the time to make yourself comfortable.

4. *Keep the working area neat.* When you free a limb, put your axe down carefully and move the limb aside. Watch where you walk, particularly when stepping backward. Don't climb over a tree with an unsheathed axe in your hand. Lay the axe down, get where you have to go, using both hands, and then reach back across the trunk for your axe.

5. *Work methodically, thoughtfully, and rhythmically.* Don't rush yourself or let anyone rush you. Stop, at least for a while, when you are getting tired; that's when accidents are likely to occur. Don't be discouraged if you get tired after working with an axe for a few minutes; endurance will come with practice.

6. *Keep both feet on the ground.* Don't stand on the trunk the first few times out, and when you do get to that point, redouble your cautionary measures.

When a trunk is raised off the ground, supported by one or more of its own limbs and possibly by its root mass, too, decisions have to be made. If you're in any doubt as to whether a particular limb is supporting the trunk, leave it alone at this stage; there will be time to take care of it once you begin cutting up the trunk. Remove only the limbs that may get in the way of cutting the trunk free from the root mass and crown.

Avoid "windfalls." When a tree has been deliberately pushed over by a bulldozer, and rests on its limbs and root mass or against other trees heaped up at the construction site, the amateur woodcutter may apply what he has learned to limb it safely. But a tree that has been toppled by the wind, whether lying alone or tangled in a pile of other trees, should be left alone by the amateur woodcutter. *The United States Forestry Service Handbook* is emphatic about the dangers here:

> Bucking windfalls is often very hazardous and should be done only by an experienced sawyer, who takes time to examine each tree to be cut for strains, breaks, binds, and the chance of falling or rolling root-wads.

When you come to a situation in which a tree is wholly off the ground, lying on or protruding from a pile of other trees, use a saw. The problem now in employing the axe is the difficulty of safe footing—not only while making the cuts, but when a large limb springs or falls clear of the trunk. Snow, ice, or wet moss on trees in the pile makes things worse, of course.

Always be careful about climbing piled-up trees or logs. Loose bark and decayed limbs have to be avoided, as well as the slippery stuff just mentioned. And in warmer weather in many areas, particularly on older piles, poisonous snakes are a possibility. If you do choose to work on such a pile, wear stout shoes or boots and keep your gauntlets on your hands. *Don't* be quiet, and in warm weather avoid putting your hands into crevices and under logs.

Never climb with a tool in your hands. From the beginning, train yourself to work this way: going up, lay the tool as far ahead of you as you can reach, then climb up to meet it, using both hands. Repeat the process as often as necessary. Coming down, leave the tool behind you, climb down a few feet, turn around, retrieve the tool, and set it down about where you are. Proceed to the bottom in the same way, one stage at a time.

If you're working with a companion, sometimes you can have the other person hand the tool to you carefully. There are times when you can hoist a tool up with rope carried up on your shoulder for that purpose.

It goes without saying, I hope, that one never *tosses* an axe or saw up, down, or in any direction! Tools can be damaged this way. Much more serious, they can injure people within range—and the range can be quite surprising.

# Cutting Up the Small Stuff on the Site

I'm going to include under this discussion of limbing the cutting up of the smaller branches of the tree, not just freeing the trunk of the largest limbs. There is a lot of good firewood and useful kindling in such material. Moreover, it should not be left lying around after your woodcutting, since in dry weather it is a fire hazard.

Cutting up the smaller stuff is best done with the light axe or a hatchet. If you don't regularly pack an axe, a hatchet is a good tool to have with you in wooded areas. Like an axe, it should be properly sheathed and carefully sharpened; and it should be handled with as much respect and discretion as the largest axe.

Hatchets have a special and quite understandable fascination for children. If you take any youngsters along with you, be sure to spell out clearly and in advance your rules for handling this tool. Explain, too, why a hatchet should never be used to scar a living tree, whose vital growing tissue, the cambium layer, lies just under the bark. This kind of information need not be given negatively. It touches on the amazing nature of living trees, described so vividly almost 100 years ago by the great American botanist, Asa Gray, in *Elements of Botany*:

> The living parts of a tree, of the exogenous kind,
> are only these: first, the rootlets at one extremity; sec-

ond, the buds and leaves of the season at the other; and third, a zone [the cambium layer] consisting of the newest wood and the newest bark, connecting the rootlets with the buds or leaves, however widely separated these may be, in the tallest trees from two to four hundred feet apart. And these parts of the tree are all renewed every year. No wonder, therefore, that trees may live so long, since they annually reproduce everything that is essential to their life and growth, and since only a very small part of their bulk is alive at once. The tree survives, but nothing now living has been so long. In it, as elsewhere, life is a transitory thing, ever abandoning the *old,* and renewed in the *young.*

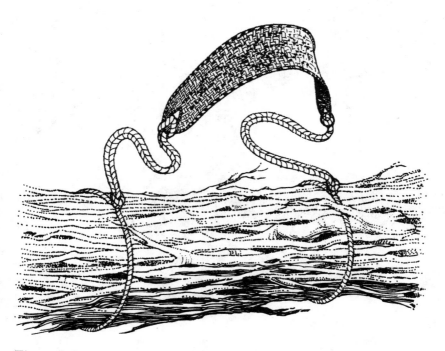

**Figure 8-6:** A tumpline, which goes across the carrier's forehead, is a kind of burden strap that can be used for transporting bundles of wood.

That trees remain alive even during the winter, when the deciduous types usually lose their leaves and look dead, is another equally wonderful tale. But it has everything to do with why a seemingly unimportant gash on a tree in winter can become an infected wound when spring returns.

**Figure 8-7:** Like this Indian woman, use a burden strap, held across the chest, to carry branches off the site.

Trimming, cutting, and tying up those valuable bundles of branches and twigs are tasks in which children can be helpful. While you're dragging your logs out with rope or chain, youngsters can carry out *their* firewood Indian style, using a tumpline or a burden strap, as shown in Figures 8-6 and 8-7.

If they would rather play, carry out the branches and twigs yourself, not only in the interest of tidiness and economy, but as a little gesture of solidarity with people elsewhere in the world who would be only too glad to have a chance to glean that litter.

Now, with limbing done as far as you can on the tree, and with branches and brush cleared away and as much of it as possible bundled up for removal, you're ready for cutting up the trunk, or "bucking."

# Chapter 9

# Bucking with the Log Saw

When your tree has been limbed, you are ready to cut the trunk into sections suitable for hauling and loading. This is called "bucking." Professional woodcutters cut trunks into standard lengths, all considerably longer than the 4-foot sections that are convenient for an amateur to maneuver to and from a pickup truck or van.

Had I but skill enough and time, I sometimes feel, I would like to do all my bucking with an axe. But I was not raised in the woods, and much as I would like to, I cannot spend all day, day after day, woodcutting; so I do a good deal of bucking with a saw.

## The Pleasures of Sawing

Sometimes of necessity, but more often for the sheer pleasure of it, I use a hand-operated crosscut saw. Alex Bealer in his fine book, *Old Ways of Working Wood,* shows that I am in good company:

In the generation since World War II, which marked the real demise of hand saws among carpenters and lumbermen, the general attitude toward hand sawing is that it is an onerous operation from which mankind has been saved because of the development and availability of power saws of different types and sizes. It is true that electrical and internal combustion engines remove the need for some effort in sawing. But while they do the job much more quickly, they also remove the pleasure and the feeling of oneness between the man and his tool. The man who understands hand saws, and uses saws specifically suited to his work and maintains them as a good saw should be maintained, will enjoy far greater pleasure, withal less production, with hand saws than with power saws. The rhythms developed, the gentle sounds, the feeling of individual accomplishment in using hand saws can become intellectual and esthetic discoveries instead of work.

Before describing the kinds of situations one encounters in bucking with the saw, and ways of dealing with them, let me deal with a problem of terminology—a double problem, in fact. The kind of saw I will be talking about is one I have used myself for many years now. It is known in forestry parlance as a one- or two-man crosscut saw (Figure 9-1). My saw has the hardwood tooth pattern, which I would recommend for general use. Both hardwood and softwood configurations can be seen in Figure 9-2.

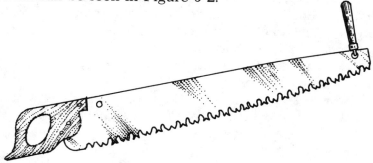

**Figure 9-1:** The one- or two-man crosscut saw.

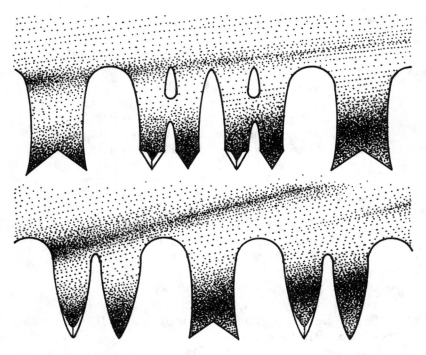

**Figure 9-2:** Different blades are used in the crosscut saw when
sawing hardwood (above) and softwood (below).

The saw is called a one- *or* two-man saw because the saw
comes with two vertical handles that can be attached as
desired to the top of the blade. One of the additional handles
may be attached close to the permanent handle. The other
may be attached (as shown in Figure 9-1) to the front of the
blade and held by a second person. The saw can then be
used as a two-man saw.

Now, besides being an ardent woodcutter, I am also a
dedicated wood*worker.* As you might suspect, I have a particular affection for hand woodworking tools, and that's
where the first part of the saw's name problem lies. In woodworking, crosscut saw refers to one of the two major types
of hand saws, the other being the rip saw.

To be sure, there is some justification for this overlap in terminology. Wood*cutting* crosscut saws and wood*working* crosscut saws are used equally for cutting across the grain. But besides the saws themselves differing substantially in size—the wood*cutting* crosscut saws are, of course, much larger—the tooth patterns differ significantly.

My excuse for dwelling on this distinction is that knowledge of all hand tools is fading fast, except among collectors. But, as Bealer's statement suggests, such tools, including saws, of course, work pleasantly and efficiently, provided you know which one to use, how to operate it, and how to maintain and sharpen it.

A well-sharpened woodcutting crosscut saw with the right tooth pattern for the trees that you are cutting through won't cut as quickly as a chain saw. But it cuts pretty fast when you know how to use it, provides excellent exercise, and is a lot cheaper to buy, feed, and maintain.

But can anybody use such a saw? How strong do you have to be? That brings me back to the second part of my terminological problem, the one- or two-*man* part. I know women can handle axes, not just from reading but from the testimony of a friend. A dignified grandmother now, she stopped and reminisced with me one day not long ago when she saw me working with my axe in our backyard. She had done a lot of work with an axe as a girl, she told me, because cutting wood for the family cook stove was an essential chore. Since her older brother was lazy, she and her sister did the cutting.

I didn't ask her whether she had worked with a saw, too, but no matter, for on this point I have the published testimony of another lady woodcutter in *We Took to the Woods.* Here is Louise Dickinson Rich's description of bucking with the saw—a two-*man* (!) crosscut saw, to be both technical and *in*accurate:

> We put up eight or ten cords of wood, all of it since the hurricane being blowdown along the Carry Road. That is not an editorial we. Gerrish and Ralph do most

of the work, but the proudest moments of my life are those occasions upon which Gerrish sidles up to me at lunchtime and mumbles, for fear of hurting Ralph's feelings, "You got time this afternoon to give me a hand? I got an old son-of-a-bitch of an old yellow birch to saw up." You see, I'm a much better hand on a two-man crosscut saw than Ralph is. Gerrish says I'm better than a lot of professional woodsmen he's worked with. This sounds like frightful bragging, but I don't care. It's really something to brag about.

Excellence on a two-man crosscut saw has nothing to do with size and strength. It's wholly a matter of method. A two-man crosscut is a saw blade four and a half or five feet long with a removable handle at each end. The sawyers take their stances at either end and pull the saw back and forth between them. That sounds easy, and it is easy if you can just remember to saw lightly, lightly, oh, so lightly. . . .

(The whole passage is worth reading, for in it, she goes on to tell quite specifically how she manages "to retain the fairy touch.")

# The Crosscut Saw

Not everyone will agree with those who like sawing for its own sake, but I have no doubt that for the amateur wood-cutter, and especially a beginner who works alone, the one- or two-man crosscut saw is a valuable acquisition. With it, you can get out a lot of wood before you become adept with the axe, which is physically harder to use and more difficult to learn to use well in different situations. If you already have a chain saw, or plan to get one soon, the hand-operated saw is a good back-up tool, insurance against a woodless trip if the chain saw lets you down.

Also, a chain saw is a little too fragile (and tempting to thieves) to keep in your vehicle all the time; there's also the problem of hard starting if you let the gas/oil mixture sit around too long. But a saw can be kept oiled and wrapped in a piece of carpeting or heavy cloth and tucked away in your vehicle, ready for use in case you come unexpectedly on some wood worth cutting.

In what follows, I'll assume you are alone and using a one- or two-man crosscut saw rather than the larger two-man saw, which must be used by two people and is shown, in case you're interested, in Figure 9-3. The bigger saw cuts faster and can handle really large trees. But my experience has been that, unfortunately, there are only too many times when it is a question of being able to cut alone, or not going out at all.

Anyway, the big decisions that have to be made when bucking are the same, whether you're working alone or with a companion, so let's look at the kinds of situations one can deal with safely and intelligently from the very beginning.

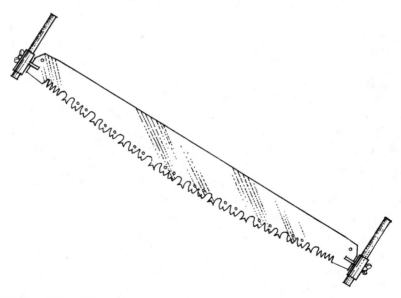

**Figure 9-3:** The two-man crosscut saw.

# A Game of Wits: Types of Bucking Situations

I'll do my best to classify the types of situations I believe most likely to occur. But I do want to emphasize that bucking is a game of wits. No two situations are ever exactly alike; each presents its own challenge. You'll find that time spent analyzing each case and planning your moves in advance will be amply repaid in safety, efficiency, and satisfaction.

Let's look again now at the basic types of positions in which you will find the trees you'll be working on (see Figure 9-4):

1. A tree may be protruding from a pile of other trees.
2. The trunk may be supported by the root mass at one end and a few heavy limbs at the other.
3. The butt, or root, end may be resting on the ground with the crown end supported by a few limbs.
4. Most of the trunk may be lying on the ground.

One general suggestion before the details—start by working only with smaller trees; no more, let's say, than 6 inches in diameter. The larger the tree, the more cutting is involved, and the greater the masses you have to contend with when things start moving.

Not only is it safer to work with smaller trees at the beginning, but also you can get more firewood by working efficiently through a number of smaller trees than by getting hung up for hours—or days—on one old giant. I haven't made many mistakes in woodcutting, but that's one I used to commit repeatedly. An eleven-year-old girl with a hatchet, working sensibly on saplings and branches, could have gotten out much more firewood than I have done on only too many occasions when I insisted on exhausting myself over some woody mastodon a couple of feet in diameter.

Now for the details:

*Situation 1.* When one end of the tree is protruding from a pile, take your saw and proceed in this way: after climbing *safely*—remember, this means never climbing

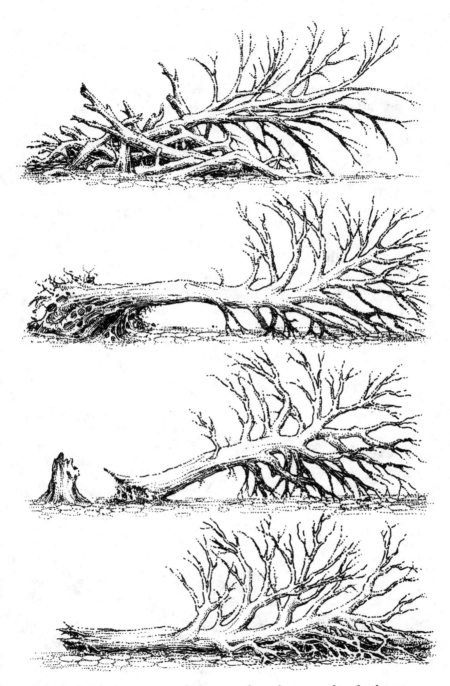

**Figure 9-4:** A tree may be protruding from a pile of other trees (top), or it can be alone, supported by its root mass and heavy limbs; it can be free from its roots but supported by its limbs; or flat on the ground, with no support from roots or limbs (bottom).

with an uncovered saw in your hand—to a secure footing for sawing, cut off the end that is protruding. Cut the trunk as close to the pile as you can without putting yourself in an unsteady cutting position or having to contend with distracting vibration from the tree itself. Do the cutting in two stages. First underbuck, that is, undercut, the trunk for about one-third of its diameter, then overbuck to meet the first cut (see Figure 9-5). You should cut in this order because if you cut all the way through from the upper surface, the trunk will probably split and splinter before you can finish cutting through.

Undercutting seems awkward at first; it's harder than overcutting, and a rank beginner is tempted to omit it. Don't. Apart from the wastage if the trunk splits, there is a real hazard. Proceeding as suggested, you will always know just when the final cut will be completed and the end is

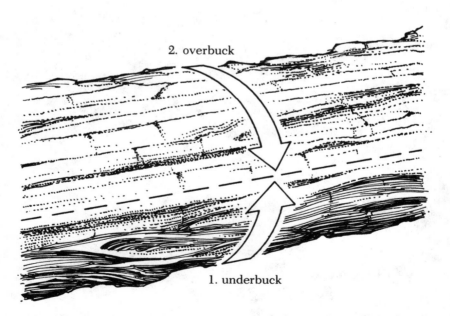

2. overbuck

1. underbuck

**Figure 9-5:**   Underbuck, then overbuck, when cutting a tree that is supported only at one end.

going to drop off. Any other way, you are taking an unnecessary chance, and the risks will increase when you go on to thicker and heavier trees.

If there are companions below, make sure everybody knows when you're ready to cut through. If you have children with you, don't rely merely on singing out. Have them sit still, in plain sight, well away from the pile you're working on, and don't let them get up until you've given an explicit all-clear signal.

Once you've dropped the end, work on down the trunk as far as you can, taking off what you judge to be the most convenient lengths.

Whether you're working on a pile or on the ground, there are always several personal safety considerations to think through *before* you begin your first cut. First, if you're working at an angle or cutting on a slope, take that into account. Whatever you cut free is going to try to drop straight down; where will you be then? Notice that I said *"try* to drop straight down." More often than not, a freed mass will not be able to come straight down. It will pivot on one or more supporting limbs, bounce a bit and swing around, or do something else that requires you to be well to the side of the cut you've just completed.

Always decide in advance, then, which area you don't want to be in when part of the tree comes down, and identify your "escape zone"—a spot where you'll be safe. This means figuring how you're going to move away quickly and making sure you will have the needed clearance and footing to move aside, perhaps backward!

Though all of this may sound a bit complicated on first reading, it's really easy for a beginner to attend to, even when working some distance off the ground—provided you are working on smaller trees and with hand tools. Get the hang of things this way first, and you will be forming safe habits for when you go on to bigger trees. It's particularly important to have this kind of experience before working with a chain saw, when all kinds of things can happen fast —too fast, in my opinion, for a beginner's safety.

*Situation 2.* Now for a situation that looks hard and is a little tricky—the trunk supported by the root mass at one end and a few heavy limbs at the other. After doing only the necessary limbing, first free the trunk close to the root mass, then separate it from the crown.

There is an important difference between this situation and the preceding one. This time you overbuck (overcut) first to a depth of about one-third the diameter of the trunk; then underbuck to meet the overbuck, as in Figure 9-6. What you have to avoid now is not splitting, but the saw being pinched as you work down from the upper surface. You avoid this by using the shallow overbuck and deep underbuck.

After you have freed the butt end, proceed to the crown end of the trunk, and cut it through, using the same se-

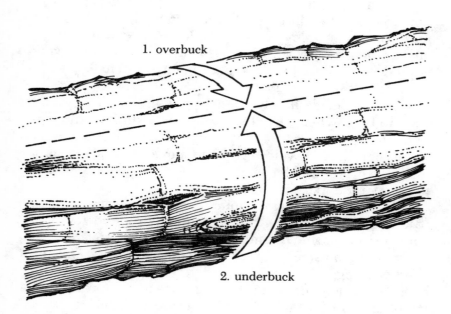

1. overbuck

2. underbuck

**Figure 9-6:**   Overbuck, then underbuck, when cutting a tree that is supported at both ends.

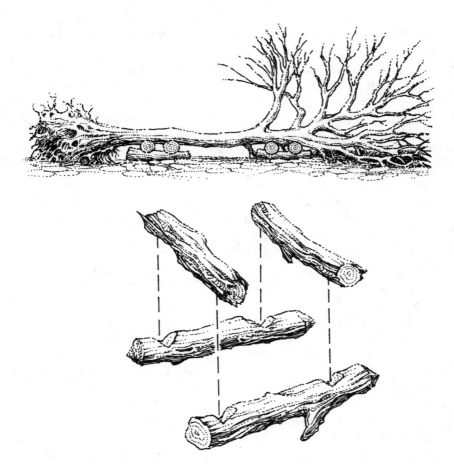

**Figure 9-7:** Blocking a tree that is supported at both ends makes bucking easier and safer. Notch out some logs with your axe to create the log crib illustrated.

quence of overbucking one-third of the way through and then underbucking to meet the overcut.

You can often greatly improve the supported-at-both-ends situation by blocking the trunk on each side a little way back from where you'll be cutting. Stones as well as logs (as in Figure 9-7) can be used. Blocking helps if there is disturbing vibration, and it is an important safety measure when working on larger trunks.

Letting the saw get caught is not only bad for the saw; you can strain a muscle pretty easily if you lose your cool and begin to wrestle with the tree to free the saw. The best thing to do if you do get your saw in a bind is to use a long pole to raise the trunk enough to spread the cut and free the saw. You can only do this, of course, on fairly light trees. For heavy trees, you will have to fetch a jack from your vehicle to raise the trunk while you shove sufficient blocking underneath. After you've done this in snow or slush, late in the afternoon, you'll decide that there are worse things than a bit of "extra" undercutting.

*Situation 3.* No need for any comment here, since the case in which the butt end is already resting on the ground is the same as the second stage of what I have already discussed in connection with Situation 2. Overbuck to one-third of the diameter, then finish by underbucking.

*Situation 4.* When most of the trunk is already lying on the ground, you have a situation, especially in the case of smaller trees, where it is easier to buck with the axe than with the saw (the next chapter describes bucking with the axe).

If you wish to use the saw, you will have to raise one end of the trunk and block it firmly high off the ground to give you room to work the saw—probably while you kneel on the ground, which is not the best position for sawing.

If you do decide to block up the trunk for sawing, be careful not to strain your back. If your regular occupation is sedentary, and if you're over thirty, be wary. You may indeed have the strength to lift the trunk easily and you may be warmed up from cutting and feel pretty good while you're doing the lifting. But you can set yourself up for some very painful inflammation the next day. Experienced woodcutters don't lift trunks or logs if they can possibly avoid doing so. They roll them with poles, levers, and other special tools, they drag them, and they hoist them with tackle. So take it easy when it comes to lifting manually even "light" logs and trunks. Better still, train yourself from the beginning to avoid such lifting altogether.

# Overbucking and Underbucking

By now, you may find all this talk about overbucking and underbucking confusing. Without carrying this book, or at least a little card, around with you, how can you remember the sequence and proportions in this or that situation? The *explanation* for what to do, and when, comes from a branch of elementary engineering theory called "the flexure of beams."

Think of a tree trunk as what it actually is, a bundle of tough and elastic fibers, something like a thick electric cable (Figure 9-8). Now think back to Situation 1, where a trunk is wedged into a pile of other trees. Because of the weight of the crown, the fibers along the upper half will be *stretched;* those along the lower half will be *compressed* (Figure 9-9). The upper half is divided from the lower half by a neutral axis along which (theoretically) the fibers are neither stretched nor compressed.

Why is it necessary to underbuck one-third of the way up before cutting through from the upper side? Because if you cut through only from the upper side, there will soon be so much tension on the few fibers still remaining above the neutral axis that they will suddenly rupture, producing the splitting and splintering I spoke about before (Figure 9-10).

In Situation 2, where the trunk is supported at both ends, the weight of the trunk itself causes its fibers to be stressed in a manner exactly opposite to the stresses in Situation 1. Now the *upper* fibers are in compression, while the *lower* fibers are stretched, or in tension (Figure 9-11). It's easy to see now, by looking at Figure 9-12, why the saw will be pinched as the cut deepens. Compression forces the

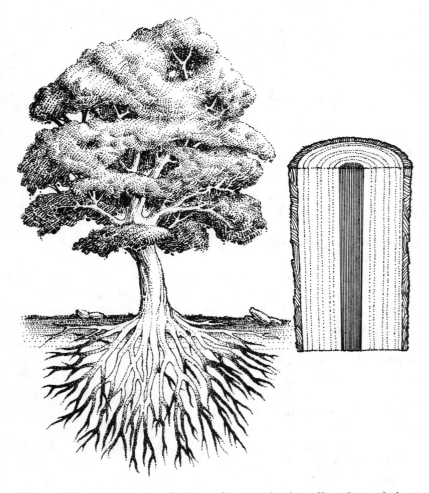

**Figure 9-8:** A tree trunk is made up of a bundle of tough but elastic fibers running vertically.

edges of the cut against the saw blade, and the resulting friction slows and finally stops the saw, just like the brakes on a wheel. We avoid this braking effect by overcutting only to a distance of one-third the diameter of the trunk, then completing the cut by undercutting.

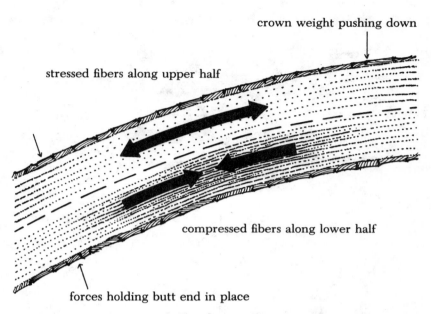

crown weight pushing down

stressed fibers along upper half

compressed fibers along lower half

forces holding butt end in place

**Figure 9-9:** The lower fibers are compressed and the upper fibers are stretched when the trunk is supported at the butt end.

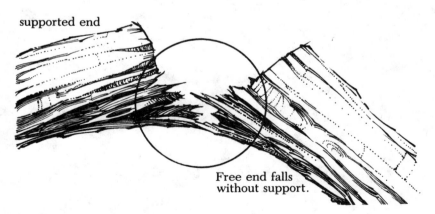

supported end

Free end falls
without support.

**Figure 9-10:** Splitting or uncontrolled shearing and separating of the fibers occurs when the stressed fibers in a limb supported at one end are cut before the compressed ones.

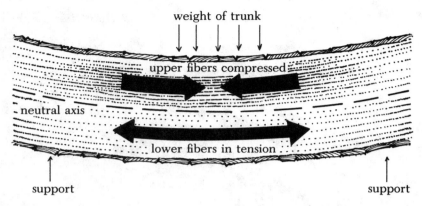

**Figure 9-11:** The upper fibers are compressed and the lower ones are stressed when the trunk is supported at both ends.

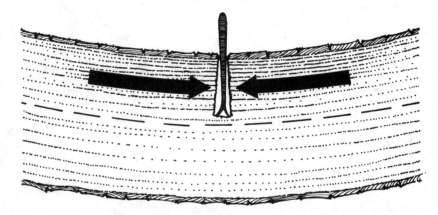

**Figure 9-12:** The saw blade gets pinched when the first cut is made into the compressed fibers of a limb supported at both ends.

# Working on Larger Trees

Starting with light trees and working thoughtfully, you will soon be ready to tackle somewhat larger trees. Additional considerations now come into play, and new techniques are required. I'll just sketch them in, because if you wait until you're ready you'll be able to work out the details for yourself. What I will be saying now applies to trees of up to a foot in diameter, which is pretty large. A lot can be done with material even larger than a foot in diameter by someone working alone without power-driven equipment; however, such effort falls, I think, under the heading of pioneering and engineering, rather than woodcutting. I have included a few references for those who (like myself) can't always resist the challenge of bigger stuff.

There's nothing effete about working with trees less than 1 foot in diameter. An oak log 12 inches in diameter and 4 feet long weighs about 140 pounds; a log 8 inches in diameter and the same length weighs about 64 pounds. That's still a respectable weight.

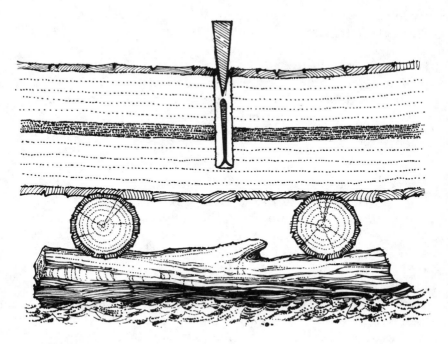

**Figure 9-13:** All really thick cuts are most practically made from the top, necessitating the use of a wedge, and in some cases, a log crib.

One of the first differences in technique when working on thicker trunks is that underbucking to any degree is not practical with the crosscut saw. All cutting is now done from the upper surface, with thin wooden wedges used to hold the cut open as the saw descends. The wedges are made in advance from seasoned white oak and driven into the cut with the poll of an axe. Blocking may be necessary to keep cut sections from dropping to the ground, and so that you can continue cutting with the saw at a convenient height. This means the construction of one or more sturdy log cribs (Figure 9-13).

Remember, too, that massive log sections cannot be dragged out, nor can they be easily loaded by hand, even if they can be rolled a short distance. If wood is to be salvaged from them, they have to be quartered on the spot, using a two-hand sledgehammer and a number of large iron wedges. Whether this is feasible depends not only on whether you can bring such equipment to where the wood is, but also on the species of tree and on the particular log you're working with. My experience has been that unless conditions are ideal, the chances of salvaging enough firewood to justify the time spent are not good.

There can be no question, of course, of using a jack on very thick logs, but there are times when you may want to do a bit of jacking on trunks and logs up to a foot or so in diameter. Use only a heavy-duty hydraulic jack, *never* a light bumper jack. Be sure there is solid footing under the jack. Work the jack only from the uphill side. Keep watching for slippage where the top of the jack presses up against the wood. As the jack lifts the wood, the center of gravity of the log or trunk will shift; the change can be dramatic, even on a slight incline, so work slowly and block up the wood before you have lifted it more than a few inches. If it needs to be raised higher, work in stages, jacking and reblocking as often as necessary. The farther you get from the ground, however, the more difficult the problem of balancing and supporting the weight becomes. Finally, never get underneath a trunk or log while jacking, and don't use the jack as a support while sawing. Once a trunk or log has been raised to where you want it, block it firmly.

If all of this sounds like a bit much for an innocent soul just looking for some firewood, I'm relieved. It means you really understand now why I'm worried about beginners tackling large trees. As you can see, the actual cutting is only part of the problem. The real difficulty is in working with the massive weights involved. A lot can be done with simple methods, but with everything around us being done by machinery today, just about all of us need to go back to the beginning, and practice on light stuff first.

Fortunately, there's plenty of relatively light stuff for the taking, and a lot of healthy enjoyment in securing it. You may want to work your way up to being able to deal manually with heavy trees using "primitive" methods. If so, you will find the passages in older physics texts on "simple machines" as fascinating as any novel. If not, there are other wonderful hand woodcutting possibilities to explore. One of them, certainly, is bucking with the axe, a topic so important and alluring that it is to have a chapter of its own.

# Chapter 10

# Bucking with the Axe

Near the end of March, 1845, I borrowed an axe and went down to the woods by Walden Pond . . .
                                    —Thoreau, *Walden*

When a fair-size tree or log is lying full length along the ground, the best hand tool for cutting it up is still the axe. To be sure, the axe as a bucking tool has been eclipsed even more than the crosscut saw by the chain saw, and detailed information on its use is hard to find. But there are a number of reasons why the amateur should consider learning to use an axe competently for bucking.

Like the one-man crosscut saw, the axe is a valuable back-up tool when working with a chain saw. It is even more portable than the crosscut saw, however, and can be carried comfortably everywhere.

Because you cut a bit more slowly with an axe, you have a better opportunity to observe yourself woodcutting. You learn more about trees and wood, terrain, and—most important—your own capabilities. This knowledge is essential for working safely in the woods.

Bucking with an axe, even more than with a crosscut saw, is a total natural exercise. When you can cut comfortably with an axe for an hour, you can be sure you're in good enough condition for any kind of work in the woods. When you first try bucking with the axe, you're likely to find that it "gets" you everywhere: in the hands, arms, shoulders, back, stomach, and legs. Sensible regular practice develops endurance as well as skill, however, so that long before you become anything like an expert, working with the axe becomes a pleasure.

City folk tend to think of the axe as a "man's tool," perhaps because traditional logging was almost completely a male occupation. But there have always been women who knew how to work well with an axe. It is by no means a brute strength implement (what genuine tool is?), and a clever, fit woman with a good axe can do all the cutting she wants to do.

Finally—and I won't be coy about this—there is the romance of the axe itself. I'm sorry that there have been *battle* axes; I insist on ignoring their existence when I think of the history of this wonderful tool. The axe has an ancient and honorable past; again and again, those who were proficient with the axe in times of peace have used it to build enduring works of craftsmanship. Look at the better examples of Scandinavian or American log construction, and you will see what I mean. Learning to use an axe well is not a step backward into the past, but a step forward, toward recovering what should not be lost.

# What Is a Good Axe?

A great deal could be said about types of axes, but there's no need to agonize over one's initial choice. Any decent full-size, 3½-pound head axe will do; but it must be correctly sharpened and properly aligned on a sound handle. (For details on hanging an axe and sharpening the edge, see Chapter 6.) It is not difficult to obtain a reasonably good axe. Reconditioned older tools can be splendid, and a number of mail-order suppliers offer good axes at fair prices. I have also seen axes at local tool stores that I would be willing to use.

I must admit, however, that I can offer no special guidance on where to buy today. Though I own quite a few full-size axes, I purchased only one of them, a lovely "Iltis Original Ox-Head Brand" in a "cedar" pattern. (Sounds like fine china, doesn't it? That's the way I feel about it!) I bought it in a large, general hardware store in Edmonton, Alberta, more than ten years ago, and have come to prize it more and more as I work with it.

I'd hate to lose my Canadian axe, but I'm fond of the others too. Each in its own way is a good tool, and each has personal associations. There is "Ira Cell's Axe," a double-bitted cruiser axe that came with an old house we bought in Nyack, New York; "Mr. Thompson's Axe," a nice Plumb axe given to me by a retired widower who had Thanksgiving dinner with us years ago; and "Granny's Axe," a 5-pound axe of considerable antiquity, with a straight handle that reaches to the top of my thigh. Although the axe is pitted with rust, the handle is solid. I put a good edge on it when I brought it home.

You become fond of axes you've used for a good number of years. You come to know the cutting edge of the blade, the fit of the handle in your grasp, and to be comfortable with the weight. But every old axe was once new, so get what you can and go bucking. Be fussy only about the condition of the tool and how you use it.

# What to Cut

The use of the axe is at once simple and endlessly complex. The starting instructions are brief and easy to understand, but you never stop learning how to use the tool properly, and the work becomes more interesting as it gets easier.

It is best to start learning bucking on a moderate-size trunk or log—at least 6 inches but not much more than 8 inches in diameter. This log should be solid enough to stand on, but not so thick that you'll get tired and discouraged before cutting through. If you're planning to cut up a log rather than a tree, and the log moves a bit when you step on it, stake it in place firmly at the ends as in Figure 10-1, or wedge it from underneath until it is steady under your feet.

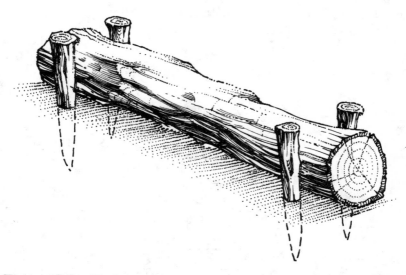

**Figure 10-1:** If you intend to buck a log with an axe, first stake it firmly in place.

# Preparing the Site

Check meticulously for interfering twigs and brush. Do this by making a few light, experimental swings, without cutting. You will be swinging the axe in a wide, high arc, so you need much more clearance now than when limbing.

Examine the ground where you'll be cutting for stones and pebbles. Take the time now to clear them away; if there are stones you can't move, keep their locations in mind when deciding just where to cut.

# Laying Out the Cuts

Before the axe begins to bite deep into the wood, you must plan the cuts. Study the trunk or log you're working on for the best places to divide it up, taking into account not only the lengths of sections you can conveniently handle, but knots and other imperfections that will make for difficult cutting. With practice, you'll be able to cut very close to

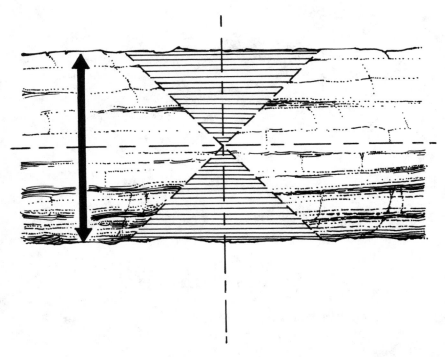

**Figure 10-2:**   Make each cut as wide as the diameter of the trunk or log.

knots and even cut through them in some species of trees; at the beginning, however, locate your cuts well away from *all* knots and hard protuberances.

The maximum width of the wedge-shaped areas you remove to divide the log should be at least as great as the diameter of the log (see Figure 10-2). It's easy to underestimate the size of the diameter, so don't guess when you are laying out the cuts. Use your axe handle as a gauge. Place the foot of the axe handle flush with the edge of the log. Place your hand on the axe handle even with the opposite edge, so that it marks off the diameter of the log. Keeping your hand in place, lay the axe handle alongside the log to give yourself the width of the cut you will make.

It is very important to give yourself enough width. Too much is better than too little, since if you start with too narrow a space to work in, the cutting gets cramped and frustrating as you get toward the center.

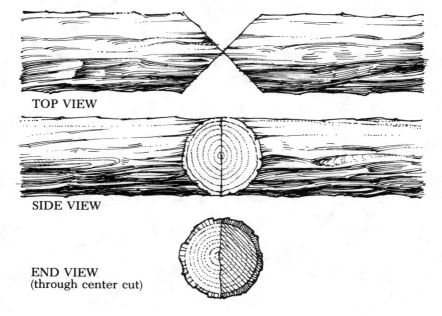

TOP VIEW

SIDE VIEW

END VIEW
(through center cut)

**Figure 10-3:** A clean cut.

Notice, too, that the faces of the cuts will be perpendicular to the ground. This is a crucial point when working on logs too heavy to be moved, since if you don't keep the faces straight up and down, the log won't separate when you reach the center. In practice, it's not hard to keep the cuts perpendicular—if you stand *on top* of the log when cutting. It may sound a bit daring, but the use of the axe as a precision tool for cutting through depends on that position; it takes much less effort to cut this way, and you can see just where the blade is to strike.

# Beavering vs. Clean Cutting

"To work like a beaver" is an expression of praise, but to say of someone that he is "beavering a tree" is another matter altogether. Here's how Ernest Thompson Seton explains the difference in his turn-of-the-century classic *Two Little Savages* (subtitled "Being the Adventures of Two Boys Who Lived as Indians and What They Learned"):

> "Beavering" was a word with a history. Axes and timber were the biggest things in the lives of the Sangerites [the story takes place in an imaginary region in Ontario]. Skill with the axe was the highest accomplishment. The old settlers used to make everything in the house out of wood, and with the axe for the only tool. It was even said that some of them used to "edge her up a bit" and shave with her on Sundays. When a father was setting his son up in life he gave him simply a good axe. The axe was the grand essential of life and work, and was supposed to be a whole outfit. Skill with the axe was general. Every man and boy was more or less expert, and did not know how expert he was till a real "greeny" came among them. There is a right way to cut for each kind of grain, and a certain proper way of felling a tree to throw it in any given direction with the minimum of labor. All these things are second nature to the Sangerite. A beaver is credited with a haphazard way of gnawing round and round a tree till somehow it

tumbles, and when a chopper deviates in the least from the correct form, the exact right cut in the exact right place, he is said to be "beavering"; therefore, while "working like a beaver" is high praise, "beavering" a tree is a term of unmeasured reproach . . .

TOP VIEW, "Beavered" Cut

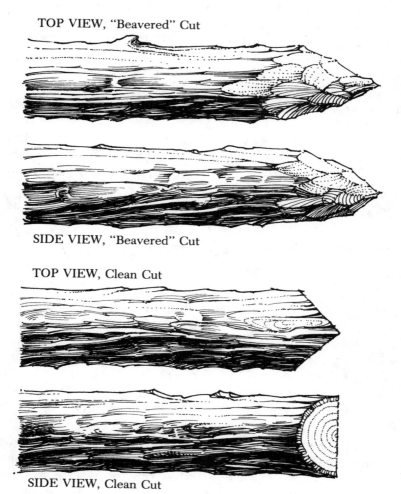

SIDE VIEW, "Beavered" Cut

TOP VIEW, Clean Cut

SIDE VIEW, Clean Cut

**Figure 10-4:** A "beavered" cut is conical and irregularly chipped away; a clean cut has smooth, regular faces tapering to a wedge-shaped end.

As a repentant former beaver, I can assure you that beavering is easy to lapse into while bucking if you don't know exactly how the cut should look when finished. With no offense intended to the clan of real beavers, let's look at Figure 10-4 to see the difference between a human beaver's cut and one that has been made properly.

Avoiding beavering is not a matter of esthetics, or a nostalgia trip. If one doesn't know that cleanly sloping wedge cuts are the goal when cutting through large stuff, bucking with the axe is likely to be a vexing and exhausting business. The cuts will be misaligned, the process of cutting through seemingly endless. You'll end up wrestling with the wood, trying to lever it around a bit for still more inartistic hacking.

I know, because I've been there, and it wasn't the light of unaided genius that showed me The True Way to Buck. Not until I found the wit to consult those old masters of woodcraft writing, Horace Kephart and Bernard S. Mason, did I discover how sensible and enjoyable bucking can be.

# Making the First Cuts

Once you know how to locate the cuts, the next step is to begin them in exactly the right way. Careful cutting, at the beginning, is necessary if the final cutting through is to proceed smoothly. Depending on the size of the log and the width of your axe blade, make one, two, or three cuts on one side of the first centerline, the same number on the other, repeating the sequence as necessary (Figure 10-5). You can see now why it's important that there be no knots or major grain irregularities in the area to be removed.

If there are no knots, and if you can place your blows accurately, a large, clean chip will soon be liberated—not wrenched, battered, or chewed—from the wood. The size and quality of the chip tell everything. Show me your chips, and I know how well you can use an axe. Look at mine on any given day, and you can tell whether I'm still making progress in acquiring the woodcutter's skill.

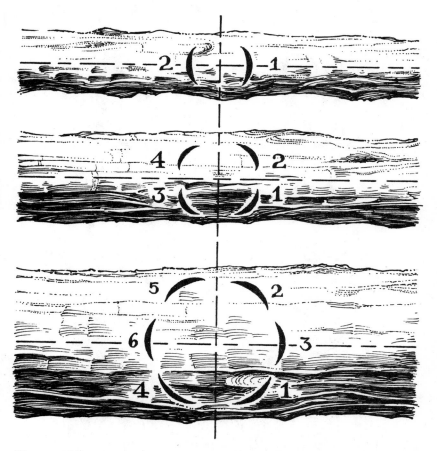

**Figure 10-5:** To buck with the axe, place cuts carefully and repeat the sequence of cuts in order until the log is cut through.

# Holding and Swinging the Axe

Now that you know what results you want to achieve, the mechanics of holding and swinging the axe need to be described briefly.

Standing on the log you'll be cutting, spread your feet considerably farther apart than shoulder width. Bend your

**Figure 10-6:**   Prepare to buck with the axe by standing on top of the log with knees bent. Hold the axe with your hands as far apart as possible. As you begin your swing, move the hand nearest the axe blade down the handle. Keep your eyes on the log as you continue the downstroke. As you finish the stroke, and make the cut, your hands should be together at the base of the axe handle.

knees slightly. At the beginning of each stroke, grasp the axe with your hands as far apart on the handle as they can go, one hand just under the head of the axe, the other at the end of the handle. When you begin the swing, hold the axe parallel to the front of your body; then smoothly and unhurriedly swing it in an arc alongside your body to a position over your head; finally, in one continuous motion, let the axe drop onto the wood where you've planned the cut. All the while, keep your eyes on the spot where you want the axe to fall. As you feel the axe descending, allow the hand close to the axe head to slide down to meet your other hand. As the axe enters the wood, both hands should be grasping the end of the handle firmly, but without straining. With practice, you will learn to finish the swing with a slight twist of your wrists, assisting slightly the splitting action of the blade.

None of these actions is difficult or awkward. Practice with a sheathed axe in slow motion, not letting the axe fall all the way, and you'll soon get the idea of what you're supposed to do.

There is, to be sure, something like a backhand swing, as distinct from a forehand swing. The difference will depend in your case on which side of your body you tend to favor. Whichever side of the notch turns out to call for your "backhand," don't attempt to shift your feet; just twist your hips slightly.

# "Good Chopping Is Gentle Chopping"

That's really all there is to it. I haven't said anything about "mastering the axe" because that puts everything in the wrong light. Bucking with the axe, like every other form of skillful hand work, is a nonviolent activity. Once you raise the axe, great force should never be applied in bringing it down. As a fair return for your work in raising it to the position from which you let it fall, gravity, recalling the

3½-pound head, does just about all the work of cutting. You need do no more than guide a well-sharpened axe into the wood at the correct angle (45 degrees, give or take a few degrees). As Bernard Mason puts it:

> *Good chopping is gentle chopping.* Never ride the ax or force it. It is the weight of the ax that chops, not the force with which it is swung. To drive the ax is hard work, and wholly unnecessary work. But worse—it destroys your aim. Accuracy is what counts in chopping. A good sharp ax will eat its way through a log quickly and cleanly if raised and dropped with a normal, natural, unforced, rhythmic swing.
>
> —Mason, *Woodcraft and Camping*

# Chapter 11

# Cutting Wood to Stove or Fireplace Length

They gave me a saw and an axe, sledge-hammer and wedges, and I spent a happy afternoon upon the hillside behind the hotel, sawing up a big log for stove wood. . . . Sunday I worked also, early and late, and Monday and Tuesday morning—and I split an amazing big pile of wood. I began to get known . . . Then logging gentlemen, between drinks, would wander up the hill to see the extraordinary person who liked work and who worked for nothing.

—Grainger, *Woodsmen of the West*

Few of us are likely to follow Grainger into the forests of British Columbia. Even if we could, it would be different, for the commercial hand logging he described in *Woodsmen of the West* disappeared years ago.

It was rough, arduous, and dangerous work. If there was miners' blood on the coal that warmed so many hearths a few generations ago, there was lumberjacks' blood on the framing and clapboards of the tidy little houses that held the hearths.

But Grainger, a Cambridge graduate who later became chief forester of British Columbia, never regretted his youthful work in the lumber camps, and he was surely right about the big and little pleasures of woodcutting that he wrote of in his splendid little book.

# The Pleasures of Backyard Bucking

Cutting up wood for burning is a minor, but by no means insignificant, pleasure. It isn't in the same league with listening to Beethoven's Ninth Symphony or wielding a double-bitted axe in a virgin fir forest, but you can do it in your own backyard on almost every winter evening. We are considering, in fact, a whole raft of pleasures; and if no single one can rightly be called glorious, who can say what the total may be worth?

There is the opportunity for regular outdoor exercise—exercise you can make as strenuous or as gentle as you choose. Except for rain and heavy snow, there is nothing to hold you back. There is no need to travel and little time has to be put in on any given day. Until the temperature drops way below zero cold is a stimulant, not a hindrance, for such work.

There is the pleasure of having something good to do out-of-doors when there is not enough time to go off to the woods—a satisfaction doubled by the knowledge that though you are working by your own back steps you are maintaining important skills for working in the woods.

Still deeper, perhaps, is the joy of an appropriate seasonal activity. Backyard bucking is the perfect cool- and cold-weather complement to gardening during the warmer half of the year. As much as we need the wood itself for fuel, we need such daily natural outdoor work to keep the less natural parts of our lives in perspective.

Outwardly, cutting wood to burning length in your own backyard is no big thing after the excitement of discovering the big stuff and bringing it in from the field. Hasty and crass types may even look on it as a chore. But we know better; far from being so much stuff that has to be cut up, those logs stacked by the house offer much more than the proverbial double heat of cutting and burning.

# Again the Axe

There are two ways of cutting logs by hand into short sections. You can use an axe, or you can use a saw. Each method has its advantages.

Let's look at the use of the axe first. Substantial logs can be bucked in your backyard, just as in the woods, by standing on top of them while cutting with the axe. As the logs get shorter, they may want to roll, but you can correct that tendency by backing the log with another log or a low chopping block cut from a piece of 10 × 8-inch timber (see Figure 11-1). Notch the block with the axe on both sides; it will then be handy for cutting wood not thick enough to stand on.

Slender pieces are cut by slanting the limb or stick against the block, one end resting against the ground, the part to be cut through placed in one of the block's notches. Cut only where the wood is solidly supported by the block; this not only makes the cutting much easier, it also prevents the freed end from flying through the air and becoming a dangerous missile. It is often necessary to squat when cutting this way, and the wood may have to be turned after every cut. A nay-sayer will call this awkward work. Think of it rather as beneficial exercise and an additional opportunity to become familiar with the axe, a tool whose usefulness keeps multiplying as you work with it in different ways.

When you are working with the axe in the backyard, a lot of chips are deposited on the ground. I like chips. The large ones, which show me that I am getting somewhere

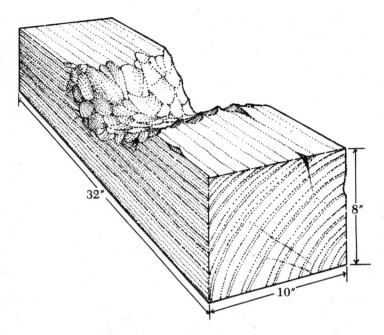

**Figure 11-1:** A low chopping block.

with my axework, are proudly displayed under the guise of being dried in the sun for fuel. This sun-drying is a very shrewd thing to do because large chips make excellent filler pieces when building up a fire.

As for little chips—I wish I could say they were few, and those only the result of getting to the narrow bottoms of perfect cuts. But the truth is that my axe-cuts, like a lot of other things I do, are rarely perfect, and the abundance of my little chips is an antidote to a vanity that swells with the growing pile of big chips.

Little chips have other uses. They, too, can be dried and sprinkled humbly on a modest fire, as befits their origin. Also, they can be used for mulch in the summer. A layer of wood chip mulch applied to the soil around well-established seedlings, trees, and shrubs will reduce weed growth, help the soil retain its moisture, insulate the plants from temperature extremes, and protect against soil erosion. Wood chips can also be a healthy addition to the compost bin. Mix them with grass clippings, leaves, and table scraps, to make an effective plant fertilizer and soil conditioner.

Even if you are as thrifty and as careful as possible, sweeping up the chips after each cutting session, there will still be a few tenacious ones that linger. You'll come on them in the dog days of summer, yards away from where you did your cutting. They, too, have their uses, as reminders that it was, in fact, cold not so many months ago and will be cold again before long.

There is still another advantage to cutting up logs into short sections with the axe. You can vary the cutting with splitting. A good many short pieces can be split with the axe alone, even the gently beveled cutting axe. Not every section will respond to this technique, of course; some won't open themselves to any axe, though there are those that may let go with a heavier, somewhat blunter beveled axe that you can keep handy for such persuasion.

# Saws

I've become fond of the axe, and employ it much more than I used to. But for years I did all my backyard cutting with saws, and I don't want to slight these reliable tools.

Notice that I said saws, not just "saw," for in addition to the bucksaw, the one-man crosscut saw has its place in backyard cutting, as in the woods. With the crosscut saw and a sawbuck you can cut hefty logs.

The sawbuck has to be solid, worthy of the logs you place on it for this kind of cutting. Fortunately, such a sawbuck is easy to make from ordinary 2 × 4 and 1 × 4 lumber. Nail it together solidly (as shown in Figure 11-2), and it will last a long time.

The only tricks worth mentioning in making such a sawbuck are to use a substantial hammer (a 1-pound head is

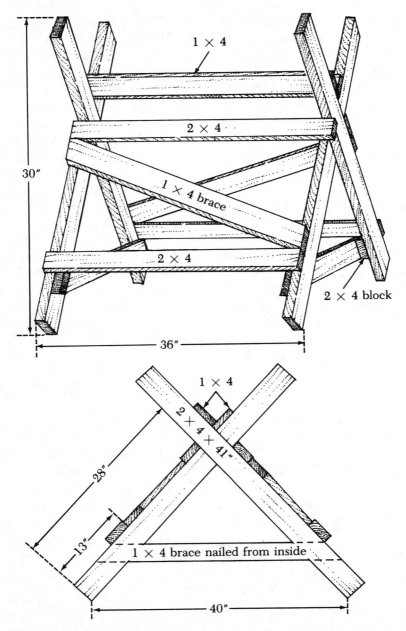

**Figure 11-2:**   Use a sawbuck to support wood to be cut to fireplace length. (Drawings not to scale.)

not a bit too heavy), and not to trim the ends of the 1 × 4 crosspieces and braces until after you have nailed them to the 2 × 4s. This will prevent splitting.

Since we're talking now about working with good-size logs—as much as 9 inches in diameter—use caution in handling them. Wear the same sturdy footwear as when working in the woods and the same gauntlets. Drag the logs to the sawbuck, using your logging chain or a length of heavy rope.

To get the log up onto the sawbuck, proceed as you did when loading it onto your vehicle. Squatting all the way down, with your back straight, lift one end of the log. Keep your back straight as you come up; your thighs, not your back, provide the leverage. Lean the upper end of the log against the trough of the sawbuck. Then, squatting and coming up as you did before, lift the other end of the log, swinging the log around as you come up and depositing it lengthwise along the sawbuck. To position the log exactly for sawing, push it along the sawbuck until one end protrudes a little more than the length of the section you want to cut off.

You may have to secure the log to the sawbuck with chain or rope. To use your logging chain, add an S-hook and a good-size turnbuckle (Figure 11-3). Rope is good, too, but it has to be heavy. I often work with a length of 1-inch-diameter Manila rope I found at a dump complete with eye splice (a loop woven into one end of the rope). Figure 11-4 shows how to secure the log to the sawbuck with rope.

Whether you are working with chain or rope, fasten the log down on the sawbuck as close as you can to the end opposite the end where you'll be cutting. Only slightly more than the section you want to cut off should protrude beyond the end of the sawbuck. With this type of rig and a sharp crosscut saw, you can cut through any log you can get up onto the sawbuck.

Another type of saw, however, is probably the best-loved backyard bucking tool. At least, it used to be before we discovered how to make a $200 gasoline-powered contraption do what a $20 muscle-powered classic can still do pleasantly. I speak, of course, of the bucksaw. It cuts through fair-size and smaller stuff easily, and requires no great

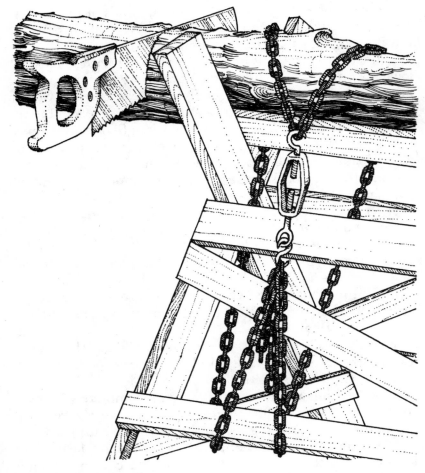

**Figure 11-3:** Tighten the turnbuckle on the chain to pull the chain taut and secure a log section on the sawbuck.

strength, agility, or skill. It is still obtainable and still relatively cheap. You can even make one yourself (more on this a bit later).

Perhaps because it is simple, inexpensive, and pleasant to work with, durable and easy to take care of—a good example, in fact, of a perfect hand tool—many people today have never seen a bucksaw, not to mention having had a chance

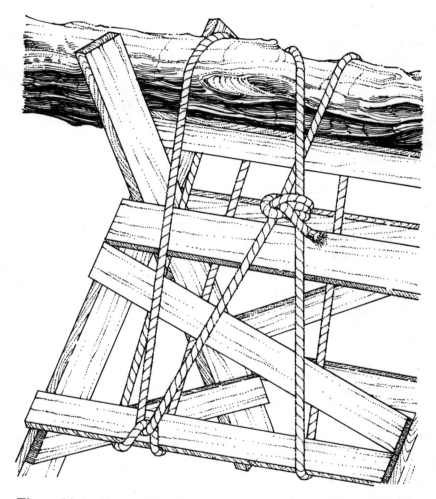

**Figure 11-4:** A rope 1 inch in diameter holds a log section on the sawbuck.

to use one. Chain saws, both gas and electric, are more profitable to market, so they are thrust on us everywhere. The bucksaw, used for well over a century when most homes fueled heating and cook stoves with wood, has to be sought for today by the canny amateur woodcutter.

To be sure, pseudobucksaws are available everywhere —wretched things with light, inadequate aluminum frames and skinny, twisty blades. These have been designed, I suspect, by people who have never sawed a log by hand. A good many people knew how once upon a time; here's an elderly New Englander talking fifteen years ago:

> The old saws was every bit as good as modern ones when it comes to doing their work. I've got an old buck saw around here now. It cuts fast—it's even fun to cut with. It has been used so much that the blade is wide on the ends and thin in the middle.
>
> —Needham and Mussey, *A Book of Country Things*

Exactly what does this fabled tool look like? And why is it so perfect for its purpose? The blade is distinctive (Figure 11-5). It is rather heavy and wide, 2 inches wide in fact. The teeth are large, $3/16$ inch long, and with a healthy set. A pin passing through a hole in each end of the blade holds the blade in the slotted handle. When the tension on the blade is released by unscrewing the nut on the threaded bolt, the pins holding the blade can be removed by hand. One side of the frame is considerably larger than the other, making a comfortable grip for one or two hands. The inward slope of the frame lowers the saw's center of gravity, placing the saw's weight over the teeth.

The threaded-bolt tension device is simple but efficient (sometimes there is a leather thong twisted with the help of a piece of wood, or a length of rope). The crosspiece is set into the frame with a mortise and tenon joint, but the tenon is not pinned or glued. Because of this design, the saw can be disassembled in a few seconds for packing or sharpening.

Technically, the bucksaw is a kind of bow saw. How many and how varied the bow saws were can be seen by glancing through Eric Sloane's *A Museum of American Tools*. Most are gone now except as collectors' items or museum specimens, but a few woodworking bow saws are still used by those who enjoy working with hand tools.

I was lucky to find one of my bucksaws under the porch of an old house I bought. Although it had been misused,

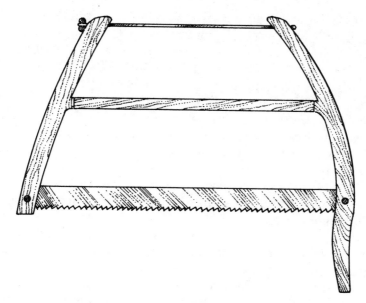

**Figure 11-5:**   A traditional wood-frame bucksaw.

broken, and cast aside, it was quite repairable, as simple tools usually are.

A simpler way of acquiring a new wood-frame bucksaw is to mail order one from a forestry or farm equipment supplier. See Chapter 5, "Tools and Equipment," for a list of companies providing this service.

If you want to be really traditional, you can buy the blade (from Woodcraft Supply Corporation) and make the rest of the saw yourself. Use any strong, fine-grained wood. The sides can be straight rather than curved, but retain the inward slope. Make the mortises less than half the depth of the width of the sides. The pins for holding the blades can be cut from large nails. Use a threaded rod, leather thong, rope, or turnbuckle for the tension device, but whatever you use, don't draw it up too tight. (Still more detailed instructions can be found in Drew Langsner's *Country Woodcraft.*)

*Good* metal-frame bucksaws have been made. I bought the one illustrated in Figure 11-6 in Canada in the late 1960s. It has a strong tubular steel frame with a lever action handle for releasing tension on the blade. The original blade was adequate, but not as wide, hence not as good, as the one shown. The blade now in the saw was cut down from a blade I found in *another* old house we used to live in. The result is a sturdy bucksaw that has kept me happy for years.

Why so much detail about such a simple tool? Because a *good* bucksaw, new or old, wood frame or steel, store bought, homemade, salvaged, or inherited, is fun to use.

True, its homely use has inspired the informal essayist rather than the poet. In *Atlantic Classics,* there is a good but anonymous essay on the pleasures of backyard bucksawing called "On Sawing Wood." Robert Frost, the American poet laureate of woodcutting, wrote a grim poem, "Out, Out," about a tragedy occasioned by a *power* bucksaw; but I didn't need that poem to decide that I wanted no such machine in my backyard. I feel the same way about splitting machines; anything capable of such violence against wood might just turn against me, too.

Who does need such machines? Whose life is too crowded for the gentle exercise and pleasant sound of sawing wood by hand in one's own backyard? The pressure of

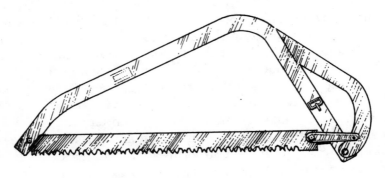

Figure 11-6:  A modern steel-frame bucksaw.

other chores, of genuine bread work is certainly a valid reason; if it's a question of being able to slump down in front of the TV a few minutes sooner. . . .

To use the bucksaw, get both hands firmly on the saw. The longer side is the one you hang on to (the teeth should point toward the shorter side). Resting one knee on the log will help you get your back into the work, and makes for much easier cutting.

As for safety reminders, only one of any consequence is needed for this kindly tool. If you start a cut with only one hand on the saw, don't rest the other on the wood close to the saw's teeth. The blade could catch, jump, and cut your hand.

Be good to a bucksaw and it will be good to you. Touch up the blade at least once a week, and sharpen it completely at least once every season.

When you hang a wood-frame bucksaw up at the end of the woodcutting season, ease off the tension a bit. Keep it with your axe and crosscut saw, out of harm's way, but not so far out of sight that they can't remind you of the happiness that lies ahead when you work with them again after nature has settled down for another long winter's nap.

# Chapter 12

# Quartering
and Splitting

> The wedge . . . works only in radial splits; such a
> split yields a collective view of all the years at once, or
> no view at all, depending on the skill with which the
> plane of the split is chosen. (If in doubt, let the section
> season for a year until a crack develops. Many a hastily
> driven wedge lies rusting in the woods, embedded in
> unsplittable cross-grain.)
>
> —Leopold, *A Sand County Almanac*

There are as many tricks to splitting wood as there are
to cutting it, and perhaps even more room for individual
judgment and style. The very question of how much to split
has to be grappled with personally, since so much depends
on where and how the wood is to be burned.

How much you decide to split will also depend on your
own experiences, bad as well as good. If you allow kindling
to run out before the end of the heating season—we've all
done it at least once, since it's surprising how much kindling

you need—some nice, seasoned fuel wood will have to be split for kindling. This is a bitter task for a thrifty soul; from it comes appreciation of the value of scrap lumber, the lighter stuff left over after limbing, and branches cut from garden shrubs.

On the brighter side, one learns with time that there are many ways of building fires, and that some fires require little or no split wood. In an open fireplace, for example, one of the best fires for a long, cold day uses thick, unsplit log sections right on the floor of the hearth (see Figure 12-1). The little "teepee" fire built up against the back logs goes to work immediately, radiating heat into the room before

**Figure 12-1:** A fire using thick, unsplit log sections provides lasting heat on a cold day.

the big wood begins to burn. Eventually the large logs catch fully, and the more obdurate they were beneath wedge or axe, the richer and more long lived the heat they now reflect back into the room. If you keep a good assortment of cut-up branches and well-dried chips, such a fire needs no split wood at all.

Some wood, perhaps a good deal, does have to be split. Depending chiefly on the diameters involved, there are two general methods of splitting wood with hand woodcutting tools. Thick sections require the use of a sledgehammer and cast steel wedges; slender pieces can be split with an axe.

# The Woodcutter's Greatest Decision— Attitude

How much wood should you quarter and split? Should you sledge-and-wedge or split with an axe? These decisions are incidental compared to the attitude with which you choose to approach the task. Learn to enjoy working with the wood in a spirit of kinship. Wood yields to the woodcutter with insight into its character more easily than it does to the one who uses only blind force.

Before you can succeed in substituting muscle power and homely implements for elaborate machinery and alien energy, you need to know why wood can be split and what happens when you attempt to split it with simple tools. Look at every piece of wood you take up to split as an opportunity to deepen your understanding of wood's properties. This makes the work of woodcutting not only easier, but continually interesting.

# Cleavability

Generally speaking, whether you can readily split a given piece of wood with hand tools depends on a number of factors. The type of tree from which the wood comes is important. Some types of wood split much more easily than others. I'm not referring to such large groupings as "pines" and "oaks," for within each of these groupings there are great differences. Live oak, for example, can be very difficult to split because of its peculiarly twisted, interlocking grain. Red oak and white oak, on the other hand, though not quite as easy to split as white pine, are usually much easier to split than live oak.

The strength and hardness of wood are not directly related to its cleavability. Seasoned yellow pine, which is wonderfully hard, strong, and durable, splits easily. Magnolia is neither strong nor hard, but does not always split easily.

The greenness or dryness of the wood can also have much to do with its cleavability. Many types of wood are dramatically easier to split when green. Oak, in all its species, is a good example of how much easier green wood can be to cut and split. With some other woods, however, the reverse is true. It's always worth experimenting, especially if you don't know for sure what you're working with.

Another important consideration is the part of the tree from which a given piece of wood comes. Knots and twisting or irregular grain configurations can cause problems. However, even in species where knots are frequent in parts of the trunk and the larger boughs, they are not necessarily present everywhere.

# Rings, Rays, and Billets

Since you will be sending your wedge or axe blade all the way through the wood, you need some insight into its structure. Wood has a marvelously complex and elegant organization. The details differ from species to species, but knowing just a few of the most common features of its structure is a great help when splitting.

You already know one of the most important aspects of wood structure: the *growth-rings* that can be traced on the ends of every clean-cut log or bough. In the case of the trunk, they are what we count to tell the tree's age. These rings (see Figure 12-2), which are concentric, mark the

vascular ray

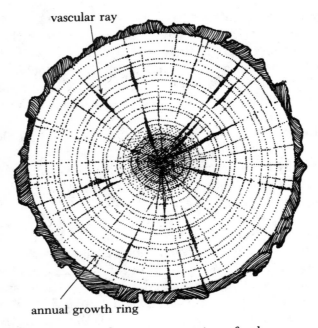

annual growth ring

**Figure 12-2:**   Rings and rays on a section of oak.

boundary between the spring wood and the relatively denser summer wood.

Besides the growth rings, there is another important structural feature visible in the end grain of clean-cut wood. This is the pattern caused by the vascular *rays,* which seem to radiate from the center. The rays, too, are visible because they differ in density from the wood around them.

The rings and rays are a major reason for wood's cleavability, because in wood, as in any solid substance, an abrupt change in density makes for greater vulnerability to fracturing from a sharp blow. In other words, the rings and rays can be used as natural lines of cleavage, paths to follow with the edge of your wedge or axe blade.

Usually, wood splits more easily along the rays than along the growth rings. This is an important help when first dividing up a thick log section. The best procedure, generally, is to divide the piece in the order shown in Figure 12-3:

1. quarters split along rays
2. subdivisions of quarters split along rays
3. billets split tangent to the growth rings

On large pieces, the quarters and radial subdivisions of quarters have to be split with a sledgehammer and wedges. The *billets* tangent to the growth rings can be split with an axe, or with a sledgehammer and wedge.

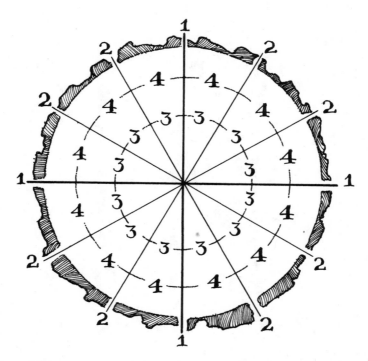

**Figure 12-3:** Logs are first quartered (1). The quarters are subdivided into radial sections (2). Finally, the radial sections are split crosswise into billets (3,4).

Besides making use of the rays when quartering and splitting, it is helpful to keep track of whether you're working from the butt (root) end of a section or the crown (branches and leaves) end. When there is a large knot, you can often do much better if you start splitting from the crown end. You then have a fair chance of running the split past the knot, or into the weak junction along its crown side. If you start from the butt end, the crack may peter out in the grain of the knot (Figure 12-4). The added difficulty of quartering and splitting wood with large knotty projections is another reason for limbing flush with the main stem in the

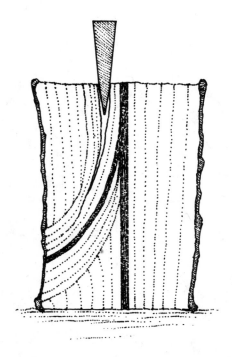

**Figure 12-4:** This shows the wrong way to split wood. When you work from the butt end of a log, knots in the grain will be likely to deflect the axe and stop the split. A knot is the result of slicing across the grain of a limb.

first place, and for care when dividing up the log with saw or axe.

# The Wedge—the Simplest Splitting Machine

You will appreciate that I have only touched on the question of what it is that makes wood cleavable—and why it is more cleavable in some directions than in others. But you know enough now for practical purposes about the wood; what about the tools you'll be using?

Simple as they appear, you can spend a lot of pleasant hours finding out what makes them work as well as they do when they are used with understanding. These simple hand tools are only relatively simple. There is a story behind them, too, and knowing even its bare outlines will help you to work much better with them.

The wedge is actually a "simple machine," as any physics text will tell you. The axe is really a version of the wedge, and the wedge itself is a variant of another simple machine, the inclined plane. If you have discovered how helpful it is to make a ramp out of two sloping logs when you are loading heavy logs onto the bed of your van or trailer, you also know *ipso facto* a valuable technical fact about the wedge: just as it is easier to move a heavy weight up a long, gently sloping ramp, so a wedge that is long in relation to its thickness is more effective as a splitting machine than a short, stubby wedge.

No matter how shrewd you are in observing the rings and rays in your wood and choosing your wedge, you know that you can't split the wood just by pushing the wedge into it with your bare hands. The wood wants to stay together, as it was supposed to do, after all, when the tree was alive.

You can get some idea of how strong this "desire" is by looking at tables of the strength properties of wood. Two convenient sources of such information are the United

States Department of Agriculture's *The Wood Handbook,* a canonical work for anyone interested in wood, and Forbes's *Forestry Handbook.*

One type of strength you're trying to overcome when splitting wood is known technically as "resistance to shear parallel to grain." In the case of white oak, it has been determined experimentally that the wood when green can stand a force of 1,500 pounds per square inch applied parallel to the grain before it shears, or separates; when well dried, however, it can take a force of 2,000 pounds.

These tests were made on small blocks of clean wood, whereas you're trying to split large hunks of wood that are usually *not* free of knots or other grain problems. In addition, other strength properties of the wood resist your efforts to split it. Can you really overcome so much strength with hand tools, even if they are simple machines?

Place the tip of a well-shaped wedge along the ray of the wood and deliver a sharp, heavy blow to the top of the wedge. A 6-pound sledgehammer, which anyone can wield comfortably, can easily deliver an impulse of 3,500 pounds to the wood, an impulse concentrated along the cutting edge of the wedge. So you can readily develop enough force to get off to a good start in splitting the wood.

Well begun may indeed be half done, and you can almost always get a crack started if you know where and how to get the wedge to enter the wood. But as the work proceeds, it's no good delivering all that force to the wood if it is used to *move* the wood instead of splitting it. You have to make sure, therefore, that the wood you're trying to split is not resting on soft ground (which would act like the padding in a football player's helmet). After all, in this situation you're trying to cause a fracture, not prevent it! You can't use stone or metal for a base because of the possibility of damaging the point of the wedge, so you must provide a solid wood base or support underneath the wood you're working on.

You now have what you need in the way of tools: a long, strong, gently tapering wedge; a striking tool that can deliver a heavy blow to the top of the wedge; and a reasonably inelastic support under the wood to be split. Satisfy these requirements and you can efficiently split with hand tools most of the wood that comes your way.

# Splitting Wood with Wedge and Sledgehammer

The general principles of wood structure and the operation of the wedge have to be taken into account by everyone who splits wood with hand tools, and there are all kinds of ways of applying them in practice.

From now on, I'll provide more description than prescription. Except for the safety advice, make your own choices as to tools and develop your own approach. When difficulties arise—and they will—go back and reflect on first principles before giving up on hand splitting. With patience, there is a hand-tool solution to just about every splitting problem. There are even solutions to the problems that can't be solved—but we'll save that discussion for the end.

Over the years, I have worked with a variety of hand tools for splitting large-diameter log sections. The list is long, but only the wedges and one of the handled tools were purchased specifically for woodcutting. Unless you're determined to split wood of every species and thickness, only a few items (which I'll specify shortly) are necessary for a beginner's splitting kit.

The handled tools I own and use are:
• splitting maul with 7-pound head
• two-hand sledgehammer with 10-pound head
• two-hand sledge with 6-pound head
• one-hand sledge with 4-pound head

My wedges were purchased one or two at a time over a period of years. By now each has its own personality for me, so permit me a few remarks along with the weights and measurements:

| Weight (in pounds) | Length (in inches) | Thickness at top (in inches) | Comments |
|---|---|---|---|
| 3 | 7½ | 1½ | These are too small for starting splits in large pieces. |
| 4½ | 8¼ | 2 | The bevel is too blunt for starting cracks. |
| 5 | 8 | 1¾ | I bought a pair of these when I got my 10-pound sledge. They are not wood-splitting wedges, but come in handy when large cracks have been started in big pieces. |
| 5 | 9 | 1¾ | These are the indispensables as far as I'm concerned— long, strong, and well shaped. When finishing up a split, I often use the 6-pounder as a hand axe, chopping away fibers in true neolithic style. |
| 6 | 10 | 2 | |

Besides the two longest wedges, the only other tool I consider essential for splitting big stuff is the 6-pound sledgehammer. The other wedges certainly come in handy, and I'm glad to have them, but the long, heavy ones do most of the work and could do all of it.

I'm fairly large (6 feet, 180 pounds) and used to working with heavy tools, so I use the 10-pound sledgehammer a lot; it does make things go a bit faster. But the 6-pound sledge can certainly deliver enough force in most situations, and it has one very important advantage over the larger sledge-hammer. The head of the 6-pounder is not much wider than the heads of the two largest wedges. Thus, if the wedge has been driven below the upper surface of the wood and the wood still won't let go, you can continue driving the wedge on down with the 6-pound sledgehammer (Figure 12-5).

The splitting maul is widely advertised as the answer to all hand woodcutter's splitting problems, and it certainly has its uses. But on tough jobs it is not an adequate substitute for a two-hand sledgehammer and a pair of large wedges. Also, I find the two-hand sledgehammer a better balanced tool. On the other hand, my journal of some years back tells me that I was happy enough with the splitting maul once, so decide for yourself. There is no disputing taste in wood split-ting, as in other matters.

Using a two-hand sledgehammer or a splitting maul properly does not require heroic strength, but rather a sense of balance and timing. The weight of the head, striking a sufficiently long, heavy, and sharp wedge, does the real work. Position the wedge intelligently (tapping it just far enough into the wood to stand alone). Spread your feet com-fortably and plant them solidly. Bend your knees a little. When raising the sledge, place your hands far apart on the handle. One hand is just under the head of the sledge. The other grasps the lower part of the handle, about 3 inches from the end. Then, as when working with the axe, the higher hand glides along the handle to meet the other hand while the head is falling. Here, too, there is no need to grip the handle tightly. Usually, you should not add a great deal of extra force by driving the head down; rather, just let the sledgehammer fall. Swing the sledgehammer smoothly in an arc alongside your body. As the head drops, keep your eyes on the top of the wedge and guide the sledgehammer gently and accurately, so that the striking face of the sledge comes down squarely on the wedge. Let the hammer rest on the wedge a second or two; then repeat the swing.

**Figure 12-5:**   A 6-pound sledgehammer drives the wedge deeper into the split.


The little one-hand sledge is designed for stone work, but is a useful addition to the woodsplitter's tools. It can be used with the heaviest wedges, and is handy when you have to hold the wood with one hand. This tool *is* driven with muscle power; you make up for the relative lightness of the head and the lesser distance it travels by increasing the speed with which it travels.

But don't overdo things, even with the little one-hand sledge. There are a number of ways to hurt yourself with these homely tools. Please treat them with as much respect as the more obviously formidable cutting tools. Always wear gauntlets and sturdy shoes when splitting. Keep the wedges in good condition; this means keeping the edges square and sharp, and filing or grinding off mushroomed tops. Keep your striking tools in good condition, too. *Never* work with a sledge or maul that is loose on its handle, or with a cracked handle. If you don't wear eyeglasses and want to be extra safe, consider wearing goggles or safety glasses.

And now for the oddest, but most important, safety rule of all. *Don't work too hard.* The lighter the handled tool you're using, the more likely you are to drive it with considerable force. This is all right up to a point, but be very careful. A glancing blow can send the head of the hammer into your shin or kneecap. It's really safer to use a heavier tool, with gravity doing most of the work of bringing the head down, than to force a lighter tool beyond your ability to keep it under control *at all times.*

Take care of yourself first and foremost, but watch out for your splitting tools, too. A new sledgehammer and a pair of good wedges are a substantial investment. Wedges don't break, but they can get mislaid. Keep them in a sturdy bag or an old pail, and don't let them stray far from the sledge. It's harder to mislay a sledgehammer than a wedge; unfortunately, it's not hard to break the handle close to the head. Watch out for friends who have heard "The Ballad of John Henry" or who want to show off their muscles. The result is likely to be a wild blow, with the wood of the handle striking the wedge. There will be no damage to the wood that was to be split, but there is a good chance that you'll end up with a cracked or broken handle.

Every now and then, you will run into a piece of wood that is just too much, at least for the moment. It is better to knock off work, even if the wedge appears to be hopelessly buried in the wood, than to get impatient or desperate. You won't solve the problem by fighting the wood; more likely, you'll hurt yourself.

Leave it alone overnight if you possibly can. The wood won't rot, the wedge won't rust away; in fact, leaving the wedge for a number of hours may of itself weaken the wood fibers sufficiently that the piece splits easily in the morning.

Sometimes the hardest thing in splitting large pieces is being able to support the wood properly. A solid piece of heavy timber makes a good low base, and a stump or large-diameter section of sound wood makes a good raised block. The latter is particularly desirable if you like to work with the one-hand sledge.

**Figure 12-6:** Support sections that have been bucked with the axe as shown here.

Sections to be split that have square ends will stand upright by themselves on a level block. But what if your sections have been cut with the axe and have chisel-shaped ends? There are at least three answers. Smaller and easier stuff can be held with one hand while you use the other to drive the one-hand sledge. Sometimes you can split the wood lengthwise, laying it in the notch of the low chopping block and driving your wedges in from the side. Big pieces, too heavy to be split with the little sledge and too thick to be split from the side, require another approach. If you wedge the piece upright, bracing it at three or more contact points, it will stay erect. The chopping block, a heavy back log, and one or two thick sections can be arranged for this purpose around the piece to be split (see Figure 12-6).

A final remark on large-diameter sections, particularly those with twisty or irregular grain: cut them as short as you can before you begin quartering them. A long, dry piece of oak, say 2 feet long and 12 inches in diameter, can resist every tool you possess. So cut the sections of your more refractory wood quite short before you commence the sledging and wedging. Remember, too, the difference in cleavability between green and dried-out wood. Making the sections short *and* quartering them while they are still green will usually make the job much easier for you.

# Splitting with the Axe

Sections of lesser diameter, and the subdivisions from quartering thick log sections, can be split in several ways. If the ends are square (cut at right angles to the axis of the log), they can be split with a splitting maul, hatchet, or hand sledge with one wedge on the stump chopping block—or they can be split with the axe.

Nothing more has to be said about the first three methods by now, but splitting with the axe is worth a little discussion. This is the best way to split pieces with pointed or irregularly shaped ends, and of course, it's also another opportunity to practice axework.

The axe used for splitting can be the one you employ for bucking in the woods and cutting logs to sections of burning length. If you're lucky enough to have another axe, especially one with a head heavier than 3½ pounds, it could be the ideal splitting tool for tougher customers. Keep the bevel on an axe used exclusively for splitting somewhat blunter than on a cutting axe. Another possibility is to work with a double-bitted axe, one edge having a shallow bevel for cutting, the other a blunter bevel for splitting.

Splitting with the axe is best done against the low block. Rest the piece to be split in the notch in the block on the opposite side from where you are standing. Strike only where the billet is solidly backed by the block. This not only prevents wasting much of the force of the blow, but also protects the axe from going through into the ground, which will dull the edge even if there are no pebbles to cause nicks.

Splitting with the axe is an excellent way to develop accuracy; for a while, though, you must expect to experience more in the way of soul-deepening humility than exaltation over your burgeoning prowess. Round pieces need to be cleaved right down the center. Flat pieces are sometimes delightfully easy to sunder; you feel great, and understand as never before the emotions behind Robert Frost's wonderful woodsplitting poem, "Two Tramps in Mud Time." But then, as soon as a few pieces have yielded to you, your axe begins to wander around in a piece of tough, twisty-grained stuff, and you realize that there is still a lot to learn.

Here are a few modest suggestions. Just as it is advisable to cut tough, twisty-grained stuff quite short before attempting to quarter it, so it is better to quarter and subdivide quarters by sledging and wedging until you have fairly thin pieces to split with the axe.

And just as it is foolish to fight the wood when working with sledge and wedges, so is it unwise to attempt to batter it into submission with the axe. Use your eyes more than your muscles, and your mind more than your glands. The twistiest piece has a grain pattern. In vertical sections, you can always find the dark centerline of the pith, or the longitudinal striations of the growth rings. If there are twists, curls, knots or other eccentricities, you can see them, too.

So, even when there are complications, there are still the natural lines of cleavage and better paths to follow. If you slow down and strike thoughtfully, lining up successive cuts purposefully, the wood will probably open itself to your axe before long. It's the same old story: restraint and insight are required, not abandonment to aggressive instincts and reliance on blind force.

# Solving the Unsolvable

But not everything will work all the time, and there may be some unsplittables too large for burning. What do you do then? Do you look on them as conquerors, give up hand splitting, and acquire a mechanical splitter?

You can *always* discover uses for such pieces. They can be used as chopping blocks or garden seats, or as examples of natural sculpture. You may regard them just as something different, interestingly independent, and therefore worth looking at from time to time.

# Chapter 13

# The Chain Saw

It has been several years now since I decided that it is much more enjoyable to cut wood with hand tools than with a chain saw. But there was a time when I was an advocate of chain saws. I still own the chain saw I bought in 1968. For years, I serviced and repaired it myself, breaking it down completely many times. One summer I repaired chain saws for a tree nursery owner. His saws were bigger and better than my saw, but there were always several in need of adjustment or repair.

In those days, the anatomy and physiology of the chain saw were intensely interesting to me. I still remember the first winter I bought my saw; while more sensible people on the commuter train relaxed with magazines and novels, I worked through Jud Purvis's *All About Small Gas Engines* and other treatises on two-cycle engines. I liked my chain saw, noisy, smelly beast that it is, and was happy to have it.

But the world turns, times change, and lovers are fickle all. I keep my chain saw in running condition for old times' sake. Its spark is still as "fat," blue, and lively as when the saw was the apple of my eye. When woodcutting time comes, however, I leave the chain saw behind in the shop while I bring my hand tools out for the fun.

# The Enchantment of the Chain Saw

I believe that hand woodcutting can be as productive for the domestic woodcutter as working with a chain saw. But in view of the chain saw's prevalence, it's just as well to know something about it; and I do have some bits of information that you're not likely to get from a salesperson or an owner's manual. First, let's look at the arguments for buying a chain saw.

"Everybody" cuts wood that way now.

A chain saw can cut a lot of wood fast.

It's easier to hold a chain saw and let it do the cutting. Why work harder when you don't have to?

Even though the initial cost is substantial, you will quickly recover your initial investment by cutting enough wood with it to save on your heating bills.

# The Chain Saw's Drawbacks

When you look closely, most of these arguments turn out to have flaws. The wood cut with the chain saw must be transported. The saw can't do that work for you, and can put you in a situation where you overtax yourself simply moving the wood that has been so easily accumulated. Because a new saw will cut rapidly through substantial logs, a beginner unaccustomed to heavy manual work is particularly subject to this undramatic, but potentially serious problem. Cutting with hand tools, however, is first-rate exercise and warms you up completely. This reduces the chances of a back injury when handling the cut wood.

There is no denying that for someone working alone a chain saw cuts much faster through thick trunks and logs. But on smaller stuff—6 inches in diameter or less—one can work quite rapidly with a hand-operated crosscut saw or with an axe. It takes more practice and physical conditioning, but once you get there you can accumulate a lot of wood in no great amount of time.

The purchase price is only part of the cost of the saw. Also to be considered are:

• interest charges if you buy on time
• gasoline and oil
• accessories specific to the saw (gas can, file, case)
• hand woodcutting tools required even when limbing and bucking are done with the chain saw
• maintenance and repair costs
• depreciation

Maintenance, repair, and depreciation are worth looking at more closely, since no one but an experienced owner or a chain saw repairman is going to tell the whole story here. The chain saw is a high-maintenance piece of machinery. The engine operates at a much higher rate of speed than an automobile engine, or even than most other two-cycle engines, for example, those in outboard motors and power lawn mowers. For ease in handling, the chain saw motor housing has to be made of lightweight alloys even though it is subject to great stresses and shocks.

To run well and safely, everything in and on the saw must be in good working order. If you don't properly maintain the chain saw, it will let you down in the woods, and you'll get no firewood on an outing or have to cut it all by hand after all.

A well-made saw can be kept running well for five years or more *if* it is used carefully and properly serviced and repaired. But it won't take care of itself. Unless you're prepared to junk the saw after a few seasons, you must either

pay an experienced mechanic for his services or learn to
work on it yourself.

The high purchase price and the need for frequent
maintenance are some of the chain saw's major drawbacks,
but not the most important one: the chain saw can be unsafe,
especially in the hands of a beginner, who is inexperienced
in the proper use of the tool. Dealers and manufacturers
aren't exactly strident on this point. So, let's hear from a
more impartial source, the Clemson University Extension
Service. In their circular, *Home Heating with Wood,* they
make this warning:

> The chainsaw is a very dangerous tool if it is
> misused. It is involved in about 30% of all woods acci-
> dents among professional loggers and is the greatest
> single hazard in woods work. A note of caution to begin-
> ners and occasional users—be sure you understand the
> proper use of the chainsaw and are willing to treat it
> with the caution and respect it deserves.
>
> Among the more important hazards associated
> with chainsaw use are:
> - Cuts by the chain on the underside of the guidebar
>   while the chain is in motion.
> - Cuts by the moving chain on the top side of the guide-
>   bar as a result of kickback.
> - Injuries from overhead by wood or branches shaken
>   loose by tree vibrations.
> - Possible hearing impairment from excessive noise.
> - Injuries from falling trees or rolling logs.
> - Falls, sprains, and strains while carrying or operating
>   saws.
> - Injuries from starting the gasoline motor.
> - Fire from exhaust sparks and from improper fueling
>   and fuel storage.
>
> Kickback is the most unpredictable of all saw haz-
> ards. It occurs when the saw bar is thrown rapidly back-
> ward in a circular motion and pivots about the opera-
> tor's hand at the saw handle. Many times this causes a
> serious cutting injury, a loss of balance, or a loss of con-
> trol of the saw. Kickback can occur when the saw chain
> is subjected to an abrupt change in wood characteris-

tics, such as hitting a knot, running the chain too slowly, or twisting the saw so that the chain grabs. When buying a saw, consider the safety features available to help prevent kickback accidents—anti-kickback chain, anti-kickback noseguards, and chain brakes.

I'll add that the *user's* safety features are even more important than the saw's. The more you know about wood, the less likely you are to allow a kickback to occur because of "abrupt changes in wood characteristics." Of course, there are such characteristics! That's how wood is structured. The surest and safest way to learn about them is by working with hand woodcutting tools. Because they cut more slowly, and there is an intimate union between you, the tool, and the wood, you get to know wood as you cannot possibly know it when a motor-driven device is interposed between you and what you are cutting.

Though I don't want to pile horror on horror, I must mention, too, that the list of chain saw hazards just quoted is not complete. The cutting chain must be in good condition and in proper tension at all times. A worn chain can snap and fly around like a whip; a loose chain can slip off the guidebar and injure the user as well as damage the saw.

Another drawback to the chain saw is obvious, I'm sure, to anyone who likes the woods and enjoys working for pleasure: the constant loud drone of the machine when it's cutting drowns out the sounds of the woods and other human voices. This diminishes the woodcutter's pleasure in working outdoors, and makes him less alert to his environment. There are good reasons why chain saws have been prohibited in many wilderness areas.

# How to Purchase a Chain Saw

Having done what I can to make someone think long and hard before acquiring a chain saw, I still want to give some advice about purchase and maintenance to the reader who has decided that a chain saw is genuinely necessary.

To buy a good chain saw, find out from local tree service people, or contractors, the name of the best independent repair shop in your area. Go there and ask for the senior mechanic's recommendations. He can suggest a saw suited to your individual needs. He will also know—and this is very important—which saws do not stand up well and are not worth buying. He may also be a dealer in a small way and have a few new saws for sale. They will not be the cheapest saws you can buy, and you'll have to pay cash, but the chain saw will have the factory warranty, the prices will be competitive with saws of comparable quality sold elsewhere, and the seller will have a personal interest in seeing that you are satisfied with the saw.

Buying a saw this way, you will know when to bring it in for regular maintenance. Even a good saw will not be reliable unless it is tuned up at least once a year, and kept clean and in A-1 condition throughout the woodcutting season. If the mechanic is not too busy and doesn't mind your hanging around, watch him while he's working on the saw and ask an occasional question. Skilled mechanics don't get that way unless they like machinery and take pride in keeping it in good condition; if you're sincerely interested, you can learn more about your saw in such a place than anywhere else.

# Maintenance

I won't go into the details of sharpening the chain and the pros and cons of self-sharpening devices on the chain saw. There are drawbacks besides the cost; ask your mechanic. Generally speaking, though you can and should touch up the chain by hand in the field after prolonged or difficult cutting, it is advisable to take it into the shop several times during the season for professional sharpening. A dull chain cuts slowly, wears faster, and puts an unnecessary strain on the whole saw. Cutting chains are expensive. How long yours will last depends on what you cut, how you cut, and how well you take care of it.

With respect to other maintenance matters, if you buy from someone who repairs saws, he'll show you on your own saw what you need to know. If you buy elsewhere, study carefully the manual that comes with the saw and follow all the recommendations. Protect the saw from being knocked around, keep it clean and well lubricated, and make sure it is always in good adjustment. If problems arise, go gently, and troubleshoot calmly for the most obvious causes (empty gas tank, loose spark plug lead, fouled plug, chain out of adjustment). Don't begin tinkering with the carburetor unless you're sure you know what you are doing, and don't let anybody who isn't knowledgeable monkey with the saw.

How about buying a used chain saw? Unless you are already an expert mechanic or can buy a chain saw from a little old lady who used it only once a week to get to church, either forget it, or pay a competent repairman to make an appraisal. It's worth knowing in advance whether you're getting a deal, somebody else's headache, or a piece of junk in need of a part that is no longer obtainable.

A chain saw is such a small contraption that you can learn to work on it yourself. It's interesting work, but there is no royal road to skill here either. Mechanically, a chain saw is indeed simpler than an automobile; but it doesn't necessarily follow that someone who can change the plugs and points on a car *and* retime the engine can immediately move to repairing a chain saw. There are important differences between the operation of a four-cycle (automobile) and a two-cycle engine; between an auto's ignition system and the magneto system of the chain saw; and between the respective fuel systems. The basic functioning of the chain saw has to be learned rather thoroughly because the chain saw mechanic usually has less available in the way of testing equipment. A backyard auto mechanic can purchase relatively inexpensive gauges and testing devices for every aspect of an auto engine's functioning; not so the amateur chain saw mechanic. Detailed technical specifications can be hard to obtain, and without such information the amateur, who does not have the "feel" and the "ear" of an experienced mechanic, can be hard put to diagnose a problem.

Few specialized tools are needed; the ones used are chiefly lighter automotive engine and ignition tools. They have to be good-quality tools, however, and a delicate touch is required when working with the light alloys employed in the chain saw's construction. It can be difficult to disassemble a saw; use too much force, and you can break an expensive casting.

You can start caring for your chain saw by doing the simple things that have to be done anyway to keep the saw running well. Unfortunately, there's a considerable jump between basic daily maintenance and the kind of disassembly necessary for a tune-up. If you are determined to learn and can't find an experienced mechanic to guide you, an evening course in small engine repair could be an excellent solution. Chain saws are more finicky than lawn mower motors and the simpler outboards, but the basic principles and service procedures are the same.

# Use

When it comes to using the chain saw, all the fundamentals of woodcutting described throughout this book apply. Besides watching out for your own safety, you have to think *for* the saw. New chains are expensive, and it only takes a second's contact with a stone to damage a chain. Never let the chain strike the ground, or anything but wood. When bucking large logs, have one or two wooden or plastic wedges in your pocket and use them to keep the saw from getting pinched in the cut.

Always use fresh fuel. The volatile elements in gasoline evaporate if stored too long, which can lead to starting trouble. Touch up the chain frequently when working on really hard wood. Be aware of your saw's limitations; certain types of seasoned oak, for example, may be too much for the average chain saw. If starting trouble occurs, find out what is wrong rather than yanking petulantly on the starter cord. Above all, never force the saw; when it is laboring and the wood is smoking, you're already well past the time to stop and take stock of the situation.

There is much more to living with a chain saw than the nattily dressed people who sell them in department stores ever consider. If only for the sake of economy and safety, you have to know more about this tool than where to point it after you pull the starter cord.

# Chapter 14

# Felling

Felling is one of the most dangerous and difficult jobs in the logging operation. . . . The skills required for felling cannot be attained by reading a few pages of text. . . .

*—Builder 3 & 2*

There is danger in felling a tree. It is a feat for experienced hands . . .

—Mason, *Woodcraft and Camping*

The felling of snags or large trees shall be by, or under the direct supervision of, a fully qualified faller. Men shall be given special training in tree and snag felling by competent instructors.

*—Forest Service Health and Safety Code*

When it comes to woodcutting, I am an amateur writing for the benefit of other amateurs. There are, to be sure, two distinct meanings of the term "amateur." An amateur can be someone who does something for pleasure, rather than as a livelihood. Such an amateur may be as skilled as a person who works for pay. The other meaning of "amateur" is opposed to "professional," and does indicate a difference in skill. To say of someone, "She's a real professional," doesn't mean that she works because she has to, but that she can be depended upon to do a good job quickly and efficiently.

Much of the time, the difference between "amateur" and "professional" is only mildly interesting, but when we are talking about felling trees, it is a life and death matter. I have cut and burned a lot of wood by now, but when it comes to felling I'm an amateur in the second sense only. You cannot learn to cut trees down safely from any written instructions, and my experience has been quite limited.

# The Hazards of Felling

It is possible, however, to explain the hazards and to point out specific circumstances under which felling should not be undertaken by anyone but professionals:

*At the risk of your life, don't consider felling a large, old tree.* With such a tree, it is difficult to determine the tree's center of gravity, and there is the possibility of decay in the main trunk and heavier limbs. All of the usual rules for felling presuppose sound wood throughout the trunk. When decay is encountered, the whole situation changes and becomes one in which quick and experienced judgment is indispensable.

*It can be a fatal mistake to fell a tree close to buildings or power lines.* Dropping a tree onto a roof can be a dangerous as well as an expensive mistake. Hitting a power line can be deadly.

*A falling tree that strikes one or more neighboring trees and becomes lodged in them poses grave danger.* It's suicidal folly to climb up to free such a tree; if you want to pull or cut it free, you had better know exactly what you're doing;

and cutting down one or more of the entangling trees is an even more tricky business.

*Felling a tree of any size or irregularity on sloping ground is asking for an accident to happen.* A tree on an incline will not fall the way it would on level ground, and the difference is not simple to calculate. Even a gentle incline can make a great difference in the velocity and direction of its fall. Top-heaviness or imbalance in the limbs complicates the situation even more.

Look where we are already! I've only touched upon size, shape, and soundness; nearness to houses, power lines, and other trees; and the slope of the ground. I haven't mentioned other matters that have to be considered, such as whether the tree leans, what the wind conditions are, and the type of ground the tree is going to fall on. An experienced faller knows how to take all these factors, and more, too, into account.

# A Lost Skill

Generations ago, youngsters could handle an axe more skillfully than most adults today. People were felling trees all over the continent. In Ernest Thompson Seton's *Two Little Savages,* Sam, the young hero's best friend, has no difficulty winning this wager: "I'll bet you five dollars I kin cut down a six-inch white pine in *two* minutes an' throw it any way I want to. You pick out the spot for me to lay it. Mark it with a stake an' I'll drive the stake." Seton gives a precise description of how Sam goes about dropping his tree; the feat is presented as admirable for a thirteen-year-old boy, but not miraculous.

Although Sam was a native of a fictitious settlement in the woods of Ontario, there were people as skilled as he at felling a tree with an axe. Horace Kephart lived among the mountain folk of North Carolina for years early in this century, acquiring the skills that made him the dean of camping and woodcraft writers. But even he says, in *Camping and Woodcraft*, "To be expert with the axe, one must have been trained to it from boyhood," and he is emphatic about the hazards of felling.

It's good to recover traditional skills to the degree that each of us can, but we can't return to our own childhood and grow up differently. I don't have such a background; I assume you don't, either. The opportunity to learn the art of felling a tree by watching an expert is gone now, for most of us, forever.

# The Sacrifices
# of the Settlers

Even when knowledge of woodlore and skill with the axe was common, felling a tree was dangerous. One day, in the slush of a vacant lot in Nyack, I found an old book, *Sixth Annual Report of the Forest, Fish, and Game Commission of the State of New York*, published in 1901. This beautifully illustrated book includes "A History of the Lumber Industry in New York," by William A. Fox. One part of Fox's history is called "A Dangerous Life":

> The life of the pioneer woodsmen or lumbermen was always beset with dangers peculiar to their work. The early town records make frequent mention of the fatal accidents which befell them. It is remarkable how often the first death in a settlement was that of some man who was killed by the falling tree which he was cutting . . .
>
> In the footnotes appended to the town histories in Hough's Gazeteer of New York there are twenty-one different instances mentioned in which the first death among the settlers was that of some man who was killed by the falling of a tree.

Out of consideration for the squeamish, I forbear quoting the account with which Fox concludes this section.

# Where to Find Out More about Felling

I'll assume I've convinced you that great caution is needed when deciding when, if ever, to engage in felling. There's no harm, however, in reading about felling. A safe situation for the experienced amateur does turn up occasionally; also, when you call in a professional, you do want to have some way of knowing whether you're dealing with someone who has indeed mastered this hazardous skill.

Here are a few sources that describe felling procedures. All are responsible in that they clearly emphasize the dangers; none makes any claim that reading alone can teach you to fell trees safely.

Horace Kephart, in *Camping and Woodcraft,* describes felling with both the axe and the saw. Kephart is particularly helpful on how to lay out and cut the notches in the trunk that determine the direction of the tree's fall. Like everyone else who knows what he is talking about, he doesn't mince words about the dangers. He warns that, "It is both difficult and dangerous for anyone but an expert to bring down a lodged tree." He has specific recommendations about where to be when the tree begins to fall: "When the tree begins to crack, step to one side. Never jump in a direction opposite that in which the tree falls. Many a man has been killed in that way."

Also good on felling with hand tools is Bernard S. Mason, another of the fine older outdoor writers. His discussion in *Woodcraft and Camping* covers generally the same ground as Kephart, but he has important pointers and anecdotes of his own.

Two United States Government publications should also be studied. *The Forest Service Health and Safety Code,* written for employees of the forest service, gives four pages of safety regulations for felling. Many of the rules are printed in capitals—but not merely for emphasis. At the beginning of the book is this statement:

CAPITALIZED TEXT indicates that someone was killed because of not observing the practice so emphasized.

*Builder 3 & 2*, a Bureau of Naval Personnel training manual unfortunately out of print, gives useful instructions, too—though as you can tell from the excerpt at the beginning of this chapter, it emphasizes that printed information alone cannot teach anyone all the details of felling trees safely.

# The Last, Most Difficult, and Most Dangerous Woodcutting Skill to Master

There you have it. Or should I say, "There you don't"? Read. Think about the advice and warnings from the professionals. If you wish, do a lot of *mental* felling. But please, make and keep a vow to stay away from any actual felling until you have mastered all the other operations in woodcutting. Then, and only then, consider tackling what are clearly safe and simple felling jobs. Remember, too, that a few successful attempts at felling prove nothing about how much you really know. If there is the slightest doubt about the safety of a particular situation, consult an expert before going ahead. The pleasures of woodcutting are many, but dangers are present, too. Make caution your companion when you go to gather and cut wood, and you will be able to enjoy it all the more as it burns down to embers, warming you on a wintry day.

# Chapter 15

# Fuel for Thought

In ancient China, the so-called "four recluses," fishermen, farmers, woodcutters and herdsmen, were held in great esteem by Taoists and Zennists. Indeed, the greatest of Zen patriarchs after Bodhidharma himself, the Sixth, Hui-neng, . . . was a woodcutter . . .

—Stryk, Ikemoto, and Takayama,
*Zen Poems of China and Japan*

I hope I have indicated that one of the riches of woodcutting is the opportunity it offers to develop a variety of interests. It has been necessary to touch on physical conditioning; the use of traditional hand tools; ropework; the nature and uses of wood; leathercraft and metalworking. We have dipped into these and other subjects, and justifiably could have gone considerably further into each.

I have tried to give enough information to get you well started; where there wasn't space to provide more details, I've done what I could to point you toward accessible sources for more information. And yet these are only some of the practical aspects of woodcutting; there are others which seem to plead for attention, too.

# Caring for Trees

You can't spend so many good hours outdoors cutting wood by hand without developing a kindly intimacy with trees and the conviction that they deserve more attention than they usually get. The basic principles of tree care are not abstruse; they ought to be taught in schools and learned by everyone, since trees are essential to life as we know it, producing the very air that we breathe.

Government foresters are eager to share their knowledge with individuals and groups of all kinds; and a wealth of easy-to-read, inexpensive literature is available from the United States Government and other public sources. Two good examples are *Color It Green with Trees: A Calendar of Activities for Home Arborists* and *Your Tree's Trouble May Be You!* Like so many government pamphlets, these are well written and attractively illustrated—and they make up-to-date information available at bargain prices.

To keep posted on new government pamphlets on tree care, browse through the *United States Government Monthly Catalog* from time to time. This publication is indexed yearly by title and subject matter. The *State Publications Monthly Checklist* is another important resource. This periodical announces the publication of pamphlets put out by your state's cooperative extension service.

In addition to the pamphlets, the United States Department of Agriculture has published a grand, library-size volume, called *Trees, The Yearbook of Agriculture 1949.* This authoritative and beautiful work offers the opportunity for a lifetime of learning about trees. Although it is out of print, many libraries receive the *Yearbook of Agriculture* series, so check with your local librarian about borrowing this book directly or through interlibrary loan.

# The Future of the Forests

The hand woodcutter needs no reminding that trees are as important a collective responsibility as the rest of life on earth. He or she knows that though trees can and ought to be a perpetually renewable resource, our trees and forests are no longer self-renewing. Throughout the Third World, a scarcity of firewood already constitutes an energy crisis more serious than anything we have yet had to deal with:

> More than one third of humanity—by and large the poorest third—still relies on wood for cooking and home heating. For many, the growing scarcity of wood, not petroleum, is *the* energy crisis. . . . Today, as fossil fuel costs soar, many Americans in forest-rich areas such as New England, are turning back to wood for home heating. However, few Americans face the desperate fuel-related financial pinch afflicting destitute multitudes elsewhere in the world. . . . When human demands for wood outstrip the growth of new trees, as they have in recent decades throughout much of the Third World, the result is economic hardship and ecological disaster, including the tragic demise of birds and other wildlife whose survival depends on trees.
>
> —Erik Eckholm,
> from "Firewood: The Poor Man's Burden,"
> *International Wildlife*

In North America and northern Europe another large-scale forest hazard, acid rain, is getting publicity:

> For a number of reasons, the dangers of acid rain have become much more acute in the last decade or so. The effects are cumulative and increase each year. The industrial emissions which are one basic cause of acid rain —mostly sulphur dioxide and nitrogen oxides—are increasing. And vast areas of Canada, northeastern U.S. and Europe are now approaching the point where even

a slight further increase in acidity may cause irreparable damage. . . .

While the effects of acid rain are felt first in lakes, which act as natural collection points, some scientists fear there may be extensive damage to forests as well. In a process described by one researcher as "premature senescence," trees exposed to acid sprays lose their leaves, wilt, and finally die. New trees may not grow to replace them. Deprived of natural cover, wildlife may flee or die. The extent of the damage to forest lands is extremely difficult to determine, but scientists find the trend worrisome. In Sweden, for example, one estimate calculates that the yield in forest products decreased by about one percent each year during the 1970s.

—Bill Dampier,
from "Now Even the Rain Is Dangerous,"
*International Wildlife*

Fortunately, the ecological news about forests is not altogether gloomy. There are positive developments, too, in many parts of the world, and bold, heartening proposals by individuals and groups in many countries.

# From the Woodcutter's Perspective

What does the amateur woodcutter, gathering wood for the home stove or fireplace, have to do with such global considerations? When you have cut trunks into logs yourself; dragged, loaded, and unloaded the logs; cut them down to burning length and split them into sections—you know something that cannot be known any other way about the value of wood and the effort needed to make it available for fuel. Such work is intensely pleasurable, satisfying in many ways; it is, however, *work,* an expenditure of effort to achieve a definite practical purpose.

After a couple of years of woodcutting, you know how big your woodpile has to be for your winter needs and you have a fair idea of how long it takes you to gather and split that much wood for fuel. Something similar is true with regard to vegetable gardening. Relatively few of us can raise all our own food, just as relatively few can grow all our own firewood, or even provide enough wood fuel with our own hands for all our heating needs. But after a few years of gardening you know with sufficient accuracy how much time will have to be devoted to tilling the land you have available, and what crops are worth your labor.

In both areas, the new perspectives gained from doing such work yourself may eventually be worth as much as the practical returns in fuel and food, important as those benefits are.

# The Lessons of Woodcutting

Hand woodcutting makes everyone a philosopher along Thoreau's lines. The muscular effort required is more enjoyable than formal exercise, and just as beneficial physically. The concentration required to cut and handle wood safely clears your mind of a thousand previously inescapable worries. And as you find that you can acquire the needed skills —and enjoy doing so—you begin to reflect about a lot of matters besides the woodpile, stove, and fireplace.

Why is it that hand woodcutting, which not only doesn't cost anything to speak of, but also saves a good deal of money, is so much more enjoyable than many of the civilized pleasures offered to us at ever stiffer prices by Madison Avenue? The simpler and more sensible pleasures are supposed to have disappeared along with kerosene lanterns, or to be available only to those who have made a complete break with the world of high technology and high-pressure commercialism.

You don't have to be raised in the back woods to buck

and quarter and split wood competently; nor do you have to be a forester to care for hand woodcutting tools well. But if you have lived most of your life in the city or suburbs, the satisfaction you feel in learning these "lost" skills and mastering them is deep and enduring.

The total cost of a complete hand woodcutting outfit can be less than that of a single chain saw; and if the simpler way takes longer, the tools last indefinitely and require no fuel or outside servicing. Learning to work with hand tools is a step toward economic independence.

As you swing a sledge down onto a wedge to split wood, it may occur to you that providing for your own fuel needs has taught you some self-reliance—an attitude the nation, as a whole, might be wise to adopt. How efficient can it be for us to rely for fuel on a complex, increasingly more expensive system of trucks, refineries, pipelines, tankers, and oil wells? What is there to be said any longer for a situation that is vulnerable to foreign political disruption, makes war more likely, and carries a constant risk of damage to oceans and shorelines? As for nuclear energy as a large-scale alternative, I suspect that most people who enjoy small-scale woodcutting and food production believe that they already have enough to contend with. It's bad enough to have to worry about existing pollution hazards without adding even greater risks and uncertainties.

Though there must be a better way, even the most ardent woodcutter knows that we can't simply turn the clock back. Pondering the *efficiency* of a properly sharpened axe and bucksaw, I am convinced that if there is any hope, it lies in a creative mixture of traditional and contemporary knowledge. Unbalanced reliance on expensive, excessively complex, and fragile systems for providing basic needs is a form of cultural immaturity that we as a people are going to have to outgrow.

Achieving such maturity, however, means taking action, as well as thought. If change is not to be brought about catastrophically or imposed by totalitarian governments, it will have to come as significant numbers of "ordinary" people rediscover the value of doing what they can for themselves in the literally vital areas of food and fuel.

# The Poetry of Woodcutting

The virtues of woodcutting have been praised by poets throughout history. An American thinks first, and rightly so, of Robert Frost. "The Ax-Helve," "Out, Out," "Paul's Wife," "Two Tramps in Mud Time," and "The Wood-Pile" come to mind in the woods and at the chopping block.

The woodcutter is celebrated in the literature of older cultures, as well. Gilgamesh cuts down trees in the most ancient of western epics. Odysseus builds the boat that takes him from Calypso's island with trees he has felled himself. *"Ting, ting* goes the woodman's ax" in the Book of Songs, the sixth century B.C. book of Chinese lyric verse. For another Chinese poet, 2,500 years later, the woodcutter is still an important figure:

> I stamp the snow to seek
> for the wood-cutter of the hill,
> But the wood-cutter of the hill
> has stamped the snow and departed.
> All the way are the tracks of his grass-shoes,
> As my search leads me
> into the depth of the pine-woods.

> —Chao Kuan-Hsiao,
> *The Penguin Book of Chinese Verse*

The hand woodcutter needs no scholarly commentary to enjoy such passages. For one who loves the work itself, any poetic expression of it, from any age or culture, is lucid and familiar.

In woodcutting, the practical becomes poetic. And the poetic turns out to be practical—much more practical than the narrow and dangerous notions of efficiency that the hand woodcutter soon learns to question. The woodcutter finds nothing obscure in such a statement by Emerson as this one:

> In the wilderness, I find something more dear and connate than in streets and villages. In the tranquil landscape, and especially in the distant line of the horizon, man beholds somewhat as beautiful as his own nature.
> The greatest delight which the fields and woods minister is the suggestion of an occult relation between man and the vegetable.
>
> —Emerson, "Nature"

In the woods and fields, through the changing seasons, the woodcutter witnesses the intimate, balanced relations of the community of living beings and the environment. For him, the "occult relation between man and the vegetable" is real. He is reminded of the bond that exists between us and plants whenever he works with wood, something that was also alive, and not all that long ago.

How long ago? And for how long? It is a humbling experience to count the growth rings on a log you have just cut through, to *see* how many years of life-work by the tree were needed to produce the wood you are preparing to burn. What right do we have to tear through the remains of so much life with alien force, or to waste what it has left behind? It seems only decent to cut through the wood by hand, and always to use it thoughtfully.

> Visualize the history of life on the planet as being compressed into a 24-hour day. . . . The recorded history of modern man does not begin until a quarter of a second before midnight.
>
> —Farb, *The Forest*

If there is to be a second day, there is going to have to

be a change from the dominant mode of the present day to a better way, involving much less wasteful, brutal, mechanical manipulation and acquisitiveness, much more perceptiveness and thoughtful personal effort. The amateur woodcutter believes that hand woodcutting, because of its modesty, is an important part of that way.

# Bibliography

Bealer, Alex W. *Old Ways of Working Wood.* Barre, Mass.: Barre Publishing Co., 1972.

Brown, Vinson. *Reading the Woods: Seeing More of Nature's Familiar Faces.* New York: Macmillan Co., Collier Books, 1973.

Colvin, Fred, and Stanley, Frank. *American Machinists' Handbook.* New York: McGraw-Hill Book Co., 1932.

Dampier, Bill. "Now Even the Rain is Dangerous." *International Wildlife,* March/April 1980, pp. 17–19.

Eckholm, Erik. "Firewood: The Poor Man's Burden." *International Wildlife,* May 1978, pp. 20–27.

Farb, Peter, ed. *The Forest,* rev. ed. New York: Time-Life Books, Life Nature Library, 1969.

Forbes, Reginald D., et al., eds. *Forestry Handbook.* New York: Wiley-Interscience, 1955.

Gay, Larry. *The Complete Book of Heating with Wood.* Charlotte, Vt.: Gardenway Publishing Co., 1974.

Grainger, M. Allerdale. *Woodsmen of the West.* 1908. Reprint. Toronto, Can.: McClelland & Stewart, New Canadian Library, 1964.

Graumont, Raoul. *Handbook of Knots.* Centerville, Md.: Cornell Maritime Press, 1945.

Gray, Asa. *Elements of Botany.* 1887. Reprint. New York: Arno Press, American Environmental Studies, 1970.

Grayson, David. *Adventures in Contentment.* 1907. Reprinted in *Adventures of David Grayson.* New York: Doubleday, Page & Co., 1925.

Greenwood, Charlotte H. *Trees of the South.* Chapel Hill, N.C.: University of North Carolina Press, 1939.

Harlow, William M. *Trees of the Eastern & Central United States & Canada.* Sante Fe, N.M.: William Gannon, 1957.

Harrar, Elwood S., and Harrar, George J. *Guide to Southern Trees,* 2d ed. New York: Dover Publications, 1962.

Hunt, Ben. *Complete How-to Book of Indiancraft.* New York: Macmillan Co., Collier Books, 1973.

————. *The Golden Book of Crafts and Hobbies.* New York: Golden Press, 1969.

Huntington, Annie Oakes. *Studies of Trees in Winter.* Boston: Knight & Millet, 1902.

Johnson, Walter. *Fuelwood and Wood Burning Stoves.* University Park, Pa.: Pennsylvania State University Cooperative Extension Service, 1979.

Kephart, Horace. *Camping and Woodcraft.* New York: Macmillan, 1948.

Langsner, Drew. *Country Woodcraft.* Emmaus, Pa.: Rodale Press, 1978.

Mason, Bernard S. *Woodcraft and Camping.* New York: Dover Publications, 1974.

Morley, Christopher. *Parnassus on Wheels.* New York: J. B. Lippincott Co., 1955.

Murray, Jim. *Weight Lifting and Progressive Resistance Exercise.* New York: John Wiley & Sons, Ronald Press Co., 1954.

*Native Trees of Canada.* Canadian Department of Forestry and Rural Development Bulletin No. 61.

Needham, Walter, and Mussey, Barrows. *A Book of Country Things.* New York: Paperback Library, 1972.

Petrides, George A. *A Field Guide to Trees and Shrubs.* Boston: Houghton Mifflin Co., 1973.

Pfadt, Robert E. *Fundamentals of Applied Entomology.* 2d ed. New York: Macmillan Co., 1971.

Purvis, Jud. *All About Small Gas Engines.* South Holland, Ill.: Goodheart-Wilcox Co., 1963.

Rich, Louise Dickinson. *We Took to the Woods.* Philadelphia: J. B. Lippincott & Co., 1942.

Roberts, Darrell L.; Sabin, Guy E.; and Cantrell, Randy. *Home Heating With Wood.* Clemson, S.C.: Clemson University Extension Service, 1978.

*Royal Canadian Air Force Exercise Plans for Physical Fitness,* rev. U.S. ed. New York: Simon & Schuster, Pocket Books, 1976.

Rutstrum, Calvin. *The New Way of the Wilderness.* New York: Macmillan Co., Collier Books, 1958.

Sargent, Charles S. *Manual of the Trees of North America.* 2 vols. Magnolia, Mass.: Peter Smith Publisher, 1962.

Seton, Ernest Thompson. *Two Little Savages.* 1903. Reprint. New York: Dover Publications, 1962.

Sloane, Eric. *A Museum of Early American Tools.* New York: Random House, Ballantine, 1973.

Tanner, William M. *Atlantic Classics: Essays and Essay Writing.* Boston: The Atlantic Monthly Press, 1918.

Trelease, William. *Winter Botany: An Identification Guide to Native and Cultivated Trees and Shrubs.* New York: Dover Publications, 1967.

Walton, Harry. *Home and Workshop Guide to Sharpening.* New York: Barnes & Noble, Everyday Handbook Series, 1974.

*(continued on page 208)*

# United States Government Publications: from Superintendent of Documents, U.S. Government Printing Office, Washington, DC 20402

*Color It Green with Trees: A Calendar of Activities for Home Arborists.* USDA, S/N 001–000–01557–5.

*Forest Service Health and Safety Code.* USDA, S/N 001–001–00–527–4.

*Subterranean Termites: Their Prevention and Control in Buildings.* USDA, S/N 001–000–03948–2.

*The Wood Handbook.* Agricultural Bulletin No. 72, Forest Products Laboratory, USDA, S/N 0100–03200.

*Your Tree's Trouble May Be You!* USDA, S/N 001–000–033–05–1.

# Index

Windfalls, dangers of, 111
Winter, identifying trees during, 19–20
*Winter Botany: An Identification Guide to Native and Cultivated Trees and Shrubs* (Trelease), 20
Wood. *See* Firewood
*Woodcraft and Camping* (Mason), 84, 193
*Wood Handbook, The,* 15, 100
Woodpile
  accessibility of, 101–2
  covering of, 98
  pests in, 99–101
Woods, salvaging wood from, 3–5
Woodshed, need for, 99
Wool rugs, fire danger from, 12
Work-conditioning, of large muscles, 25
Wrench, tire, need for, 44